Econometric Models of
World Agricultural
Commodity Markets

Econometric Models of World Agricultural Commodity Markets

Cocoa, Coffee, Tea, Wool, Cotton, Sugar, Wheat, Rice

F. Gerard Adams
Jere R. Behrman
University of Pennsylvania

Ballinger Publishing Company • Cambridge, Massachusetts
A Subsidiary of J.B. Lippincott Company

This book is printed on recycled paper.

International Standard Book Number: 0-88410-290-4

Library of Congress Catalog Card Number: 76-3624

Printed in the United States of America

Library of Congress Cataloging in Publication Data

Adams, Francis Gerard, 1929–
  Econometric models of world agricultural commodity markets.

  Bibliography: p.
  Includes index.
  1. Produce trade—Mathematical models. 2. Field crops—Mathematical models. 3. Markets—Mathematical models. 4. Commerce—Mathematical models.
I. Behrman, Jere R., joint author. II. Title.
HD9030.4.A32   338.1'7'300184   76-3624
  ISBN 0-88410-290-4

**To Our Fathers
Walter W. Adams
and Robert W. Behrman**

Acknowledgements

This volume is derived from research work supported by the United Nations Conference on Trade and Development (UNCTAD) summarized in a research report by the authors, "Seven Models of International Commodity Markets" (Philadelphia: University of Pennsylvania, Department of Economics, December 1974).

The authors would like to express their appreciation to L.R. Klein, V.K. Sastry, and Julian Gomez for much encouragement and many helpful suggestions. Nancy Blossom provided efficient research assistance. Successive versions of the manuscript were ably typed by Nancy Anuzis and Lynn Costello.

Contents

List of Figures and Tables

Chapter 1

Commodity Market Models in the World Economy

Fluctuations in the prices and outputs of many primary commodities—fuel, foods, nonfood agricultural products, metals and minerals—have had important, almost world-shaking, impacts in recent years. The surge of commodity prices in 1973–1974, a primary cause of worldwide inflation, produced the most severe recession of the postwar period. This economic slow-down, which caused massive unemployment in the industrial countries, extended into developing areas and the centrally planned block. The increase in the costs of fuel, fertilizer, and staple foods has endangered development plans—even in some cases simply preventing starvation—in the countries of the so-called 'fourth world.'[1] In contrast, some primary producing countries have gained enormously from the improvement of their terms of trade—the increase in oil prices particularly. The experience of recent years has demonstrated how much the world economies depend on the situation in primary commodity markets.

Yet the markets for major commodities are beyond the control of any single country. World commodity markets transcend national boundaries. Demand for fuels, foodstuffs, and industrial inputs is broadly based among many consuming areas, many of which are not consistently able to meet their domestic requirements.[2] While supply tends to be somewhat more geographically concentrated than demand, many commodity prices are determined in worldwide, substantially competitive markets. Other commodity prices are dominated by cartels or producer country blocks, an increasingly important phenomenon whose actions have important consequences for all consuming countries. Commodity market policy clearly calls for a broad, international perspective.

The wide swings in demand, supply, and price for many commodity markets are characteristics of their structure and operation. The business cycle in industrial countries accounts for variations in demand for materials which serve as industrial inputs. Weather and harvest conditions cause unanticipated changes in agricultural supply. Technological developments allow substitution between natural commodities and manmade substitutes. Long adjustment periods in supply and demand introduce cyclical characteristics into the dynamics of many markets. And speculation frequently increases the volatility of price movements. Past efforts to stabilize (or to raise) commodity prices have sometimes involved important interventions in these markets, such as buffer stocks, production quotas, etc. But these schemes have not always met with notable success.[3] The issue of price stabilization once again poses major policy questions at the present time.

Our understanding of the structure and operational characteristics of primary commodity markets has not kept pace with recognition of the need to control or regulate them. After many years of comparative neglect, economists and policymakers are now turning their attention toward world commodity markets. Knowledge in general, and quantitative information in particular, on the determination of demand, supply, and price in many markets is quite limited. We have little understanding, moreover, of the relationships between the various markets and between them and the producer and consumer economies.

Econometric modeling of commodity markets has recently emerged as an effective new approach in this field.[4] Models serve as a means for better understanding the structure and parameters of the behavioral relationships underlying commodity markets. At the same time, the models provide simulation instruments which can be used for analyzing market properties. They can be used for prediction and policy studies—for example, for testing the operation of stabilization schemes under alternative assumed conditions. Commodity market models can be the instrument to link the commodity markets to models of the producer and consumer countries. They are thus an essential element in a program to model the world economy.[5]

The objective of this volume is to increase our understanding of the principal world markets for agricultural products and to provide instruments for their analysis. Our aim was to build a set of econometric models which can be used for prediction and policy stimulation, and which may ultimately be integrated with other models into an econometric description of the world economy.[6]

There is little doubt, in view of the importance of the commodity markets and their influence on the world economy, that in-depth studies of the world's principal commodity markets are needed. But such work calls for substantial resources and a long period of preparation and research. In order to limit the scope of the present effort, we have approached commodity markets from a less complex point of view. We seek to describe the markets by constructing and operating relatively simple *but structural* econometric models. Simple models cannot, of course, yield a detailed picture, but we hope that they will catch the essentials of the mechanism of demand, supply, and price determination. We expected that the models will be useful to project commodity price response at least in approximate terms and that they would serve as a mechanism for introducing commodity market developments into models of the developed and less developed countries. They can serve as a starting point for studies of stabilization policy alternatives. It is appropriate to note here that we are reasonably satisfied with the performance of the commodity market models and that efforts to integrate them into other model systems are now under way.

Econometric models have been constructed for eight principal agricultural commodity markets—cocoa, coffee, tea, wool, cotton, sugar, wheat, and rice. These commodities are representative of the principal agricultural commodities in international trade. Included are products of significantly different characteristics—the tropical beverages which grow on trees and bushes; the two principal fiber crops, which involve very different conditions of production; sugar, which is a temperate crop as well as a tropical product; and two field crops. No industrial materials such as the nonferrous metals or natural rubber are represented in this study, in part because they have been the subject of other work carried out at the University of Pennsylvania.[7]

In each case we are concerned with explanations for supply, demand, and price. Analysis of the world market and of the available statistical material lies behind empirical estimation of the behavioral equations of each of the model systems. This analysis is presented in Chapter Three. Studies of the behavior of the models as simultaneous dynamic systems programmed for computer simulation are shown in Chapter Four and their response to exogenous shocks—so-called multiplier properties—are considered in Chapter Five. Chapter Six summarizes our conclusions about the characteristics of the commodity markets and about the behavioral properties of the models. These conclusions are rich and plentiful and offer promise for useful applications of the models.

NOTES

1. The "third world" no longer includes only poor countries. The more recent term "fourth world" designates the poor developing countries in contrast to those which have gained sudden wealth.

2. Recent wheat purchases by the Soviet Union illustrate how sensitive world markets are even to the purchase of supplementary supplies when the harvest has been insufficient to meet requirements.

3. For recent discussions of commodity agreements and control policy, see Law [1975] and Brown [1975].

4. An extensive econometric model-building tradition for commodity markets has emerged, but for any particular commodity only relatively simple experimental model structures have been built and applied. For discussions of the state of the art in commodity model building see Labys (1973) and (1975).

5. For a discussion of the LINK system, which includes simultaneously econometric models for major developed and developing regions, see Ball (1973). Adams (1973a) provides an illustration of linkage between a commodity market model for copper and Project LINK.

6. These were the objectives of the original UNCTAD-sponsored studies on which this volume is based.

7. See, for example, Adams (1972), (1973a), (1973b) and (1974), and Behrman (1971). Studies of numerous other commodities have been carried out under the direction of George R. Schink by a University of Pennsylvania affiliate, Wharton Econometric Forecasting Associates, Inc.

 *Chapter 2*

General Specification of the Commodity Market Models

Commodity markets are highly complex and diverse. Commodities are characterized by specialized conditions of production, transport and marketing arrangements, demand responses, and market interventions by various governmental and intergovernmental bodies. The specialized nature of each commodity market means, on one hand, that a great deal of detailed market information must underlie the structure of the market model and must be brought to bear when the model is used for forecasting and simulation. On the other hand, full allowance for all specialized market conditions may stand in the way of recognizing the behavioral generalities which are the basis for econometric models.[1] Detail and specialization for each of the models is, moreover, burdensome in terms of data development and research effort and would have entailed in this study the cost of being able to build and manage only a smaller number of models. With the aim of providing broad commodity coverage, it was our intent to explore whether relatively simple standardized models can capture the major features of international commodity markets. Consequently, this study has been organized around a common model specification that is applied with appropriate modifications to each of the commodities considered.

The general specification used for each of the commodities is as follows: (1) supply relations for the developed market economies, the developing economies, and the centrally planned economies; (2) demand relations for these same three country groups; (3) a

5

world inventory relation; and (4) a world price determination relation. Each model, thus, has a maximum of eight equations (in some cases there are less than eight because the commodity is not produced in all three regions).

Supply and demand are estimated for each of the three groups of countries separately because of hypothesized differences in the structural relations (which seem to be supported by the estimates presented in the next chapter) and because of the desire to be able to identify differential patterns among these three groups.

SUPPLY

The theory underlying the supply side is the traditional agricultural supply response to price. The general specification of supply relations depends on expected profit maximization subject to given production functions, historical prices, and expected weather conditions. Expected profits (Π^e) depend on expected prices of the commodity in question (P^e), the expected level of output (PRO^e), the price of variable input (P^v), the level of variable inputs (V), the price of fixed inputs (P^f) and the level of fixed inputs (F):[2]

$$\Pi^e = P^e (PRO)^e - P^v V - P^f F \qquad (2\text{-}1)$$

Actual supply or production (PRO) is assumed to depend upon a log-linear production function[3] with inputs of variable factors (V), fixed factors (F), a time trend to represent secular shifts due to technological change, development of supporting infrastructure, etc. (T), weather (W), and other disturbances (u):

$$PRO = a_0 V^{a_1} F^{a_2} e^{a_3 T + a_4 W + u} \qquad (2\text{-}2)$$

For the expected supply, the weather index and the other disturbances are assumed to have their expected value of zero:

$$PRO^e = a_0 V^{a_1} F^{a_2} e^{a_3 T} \qquad (2\text{-}3)$$

Substitution of PRO^e into relation (2-1) and the first order condition with respect to the variable factor (assuming that all prices are given and that the second order conditions are satisfied) gives the profit-maximizing variable input:

$$V = \frac{a_1 P^e PRO^e}{P^v} \qquad (2\text{-}4)$$

Substitution of this level of V into the production function in relation (2-2) and solution for the actual supply leads to

$$PRO = a_0^{1/(1-a_1)} \left(a_1 \frac{P^e}{P_v} \right)^{a_1/(1-a_1)} F^{a_2/(1-a_1)} e^{a_3(1-a_1)T + a_4 W + u} \qquad (2\text{-}5)$$

The size of the capital stock is assumed to depend on the expected product price to cost of capital ratio which prevailed i periods earlier, where i depends on the gestation period necessary for the relevant capital—i.e., the time between planting and mature bearing for tree crops:

$$F = a_5 \left(\frac{P}{P^f} \right)^e_{-i} \qquad (2\text{-}6)$$

Substitution of this expression into (2-5) gives the relation which has been estimated for each of the relevant commodity supply regions under the assumption that the disturbance term is distributed so that least square estimates are appropriate.[4]

$$PRO = b_0 \left(\frac{P^e}{P^v} \right)^{b_1} \left(\frac{P}{P^f} \right)^e_{-i} e^{b_2 \, b_3 T + b_4 W + u} \qquad (2\text{-}7)$$

To estimate this relation in terms of readily available variables, however, several additional assumptions are required.

1. The relevant product to input price ratios are assumed to move proportionately to the ratio of the UNCTAD export price index for the commodity to the OECD (Organisation for Economic Cooperation and Development) GDP (Gross Domestic Product) deflator (*PDF*). This is a heroic assumption. In the developed countries cost variables could be well represented by the deflator, but this is less

likely in the developing and centrally planned countries. In the real world there are often substantial policy, transportation, and information barriers between the world commodity prices and the prices received by producers. The more constant the impact of such barriers on product relative to factor prices, the better is this assumption.[5]

2. The formation of price expectations and the existence of adjustment processes can be represented satisfactorily by combinations of polynomial distributed lags of the historical price levels and a geometric (in the logarithms) adjustment process for actual supply. Thus we assume that price expectations are formed as a weighted average of observed prices at various times in the past. This assumption may, of course, be violated. It represents a particular problem at times of speculative price change as in 1973–1974.

3. In addition, there are lags which can be accounted for by introduction of lagged variables. It takes many years for tree and bush crops, for example, to obtain the benefits of expanded acreage.

4. Supply response to price need not be distinguished between response in the form of plantings as distinct from greater intensity of cultivation of existing acreage. Most of the response is likely to be in plantings but there may be some short run response through greater application of fertilizer and insecticides.

5. In the aggregate, technological change can be described by a time trend.

6. Dummy variables can be used to capture extraordinary weather conditions (or other sources of large random fluctuations) since the construction of weather indexes based on meteorological data is beyond the scope of this study.[6]

Finally, for the estimates presented below, some additional modifications are made when special conditions are hypothesized to be significant for particular products. The relevance of the United States' land allocations for cotton, for example, is explored in the supply relation for that crop in the developed, nonsocialist countries.

DEMAND

On the demand side, traditionally formulated equations link demand to income and price. The general specification for demand posits that per capita demand (D/POP) is a log-linear function of relative prices (PDF), per capita permanent income or product ($GDP/POP)^p$, and a disturbance term (v):[7]

$$\frac{D}{POP} = c_0 PDF^{c_1}((GDP/POP)^p)^{c_2} e^v \qquad\qquad (2\text{-}8)$$

Combinations of a geometric distributed lag for the dependent variable and polynomial lags in the right-hand side variables are used to represent the adjustments to the relative price term and to the permanent income considerations. In some cases first differences of the logarithms of these variables also are utilized to represent part of the expectation formation process. The resulting lag structure for the price responses in many cases means that the demand relations are recursive within the commodity market model.

As on the supply side, the price ratio which is utilized is the ratio of the UNCTAD export price index for the commodity to the OECD GDP deflator. This ratio probably is more satisfactory on the demand side than on the supply side since the OECD deflator captures variations in other inputs in the demanding countries more successfully than it reflects variations in factor prices in the supplying countries. Nevertheless this price ratio has its shortcomings for at least two reasons: (1) In some cases specific substitutes are important for which the variations in price are not highly correlated with the fluctuations in the OECD deflator. In such cases the price index for these substitutes is used (e.g. the price index of synthetics for cotton and wool). (2) In a number of cases the relative trade barriers have changed over time so that internal price ratios in demanding regions may not be highly correlated with the ratio used. The construction of appropriate internal price ratios is far beyond the scope of the study. However the attempt is made to represent one related phenomenon. In some cases when domestic production is low for a particular commodity, trade barriers have been kept high or increased to limit the expansion of imports due to foreign exchange constraints. Of course, the price ratio used will not reflect this phenomenon, but domestic production is included to test whether or not it has a significant coefficient because of this relation between foreign trade policy and domestic price.

For a number of data cells there are apparently not unimportant measurement errors in the series for the commodity variables. For wool, for example, the series for consumption is calculated by adding net imports to domestic production and subtracting the estimated stock change. If measurement errors in the stock variables are not highly serially correlated, a large measurement error during one year in the stock variable is not likely to be followed by so large an error in the same direction during the next year. Under this assumption, therefore, the stock level at the end of the previous period may have a significantly negative coefficient if included in the estimated demand relationship.

INVENTORY CHANGE

Inadequate data preclude the treatment of stocks on any but a global level. The available data often are inconsistent, moreover, with production and demand estimates. For most of the commodities, therefore, the inventory data series utilized is generated by taking a benchmark estimate for whatever year the data seem to be most reliable and adjusting for other years for the discrepancies between production and demand. Within the simulation model, inventory levels at year end (STK) are determined by the same procedure:

$$STK = STK_{-1} + PRO - D \qquad (2-9)$$

This procedure, together with the price determination relation, does imply that desired inventories depend on total activity levels (as measured by world demand). It does not, however, very satisfactorily incorporate speculative inventory behavior.[8]

PRICE DETERMINATION

Within the model only one price—the UNCTAD export price index—is included for each commodity.[9] The price determination mechanism takes into account the relationship between available inventory stocks and the level of demand. The primary determinant of the deflated price (PDF) is hypothesized to be the level of inventories relative to world demand (STK/D) in a log-linear function (where w is a disturbance term), with a secular time trend:

$$PDF = d_0 \left(\frac{STK}{D} \right)^{d_1} e^{d_2 T + w} \qquad (2-10)$$

Because of adjustment patterns, the right-hand side variable may enter in with a lag over one or more years or in difference form, and/or the left-hand side variable may be posited to adjust in a geometric distributed lag. Relation (2–11) also is estimated in alternative forms to test, for example, whether or not the same exponent is appropriate for both inventories and demand, or whether there are nonlinearities.

This relationship can be rewritten as a demand for inventories relationship:

$$STK = \left(\frac{PDFe^{-d_2 T - w}}{d_0} \right)^{1/d_1} D \qquad (2\text{--}11)$$

It implies that inventories are proportional to the level of demand,[10] but that the proportionality factor depends on the relative price, the secular trend, and the disturbance term. The costs of holding inventories are not well represented. To the extent that the distributed lags in the model are consistent with the formation of expectations in relative price changes, however, some aspects of speculative inventory behavior may be captured.

For some of the commodities considered below, changes in international market conditions due to government purchases or provisions seem to have effects beyond those associated with the same change in demand or supplies from private sources. This seems to be the case because such policies create expectations about future related policies. Two examples would be the large grain purchases by the socialist economies in the early 1970s and the PL 480 (and other related) distribution programs of the United States. In cases in which such policies may have been important, a quantitative measure of their extent relative to total world demand is included in the estimated version of relation (2–10).

CONCLUDING COMMENTS ON MODEL SPECIFICATION

As we have indicated previously, this relatively simple structural specification is subject to modification as required for each commodity.

Particular attention has been focused in the estimation on catching the appropriate lag structure which may in many instances reflect the special characteristics of the commodities. On the supply side this involves attention on the natural lags involved in developing new production and on the demand side it involves the time lags in patterns of consumption or of substitution between natural commodities and manmade substitutes.

We should also recognize at this point that the model specification does not allow for the very considerable government interventions which have affected commodity markets in the past and which may be even more important in the future. It has been implicitly assumed that free market forces are dominant in these markets and that the statistics on which the models are based reflect these forces fairly. In any case, this is a simplifying assumption intended to ease the job of constructing the models, and particularly to facilitate their use as a part of large systems. Much of the evidence of the performance of market stabilization schemes suggests that they have succeeded only in modifying the underlying market forces and that, except for some specific instances, the assumption of substantially free markets is appropriate.

NOTES

1. The issue of detailed realism versus simplified generalization is frequently an unfortunate barrier to communication between the model builder and the model user in the real world.

2. The distinction between variable and fixed factors is most clear for the tree crops, for which the stock of bearing trees can be increased only after a gestation period of several years. The land devoted to some other commodities (e.g., Asian wet rice) also may not be easily shifted among products due to the nature of such inputs as irrigation systems. Variable factors are those which are assumed to be variable within the growing season. Of course, even among such factors the extent of variation possible differs substantially. For example, seeds are committed at the time of planting, but the labor input may be varied until the product is finished (e.g., the decision may be made that the expected return does not merit the labor cost of bringing harvested crops in from the fields). Such differences in the degree of flexibility among variable factors are ignored in this analysis.

3. The log-linear production function is chosen because of its simplicity, desirable properties, and familiarity. Hypotheses concerning the magnitude of the elasticity of substitution among factors have not been tested for the commodities under examination.

4. The definition of the b_js in terms of the a_js should be clear by comparison of relations (2–5) and (2–7)

5. For some factors (e.g. unskilled labor), however, these barriers are sufficiently large that they effectively are not traded internationally. In such cases the necessary assumption may be quite strong.

6. Bateman (1965a) and Oury (1966), for example, have constructed such indexes for other studies.

7. For discussion of the preference functions implied by this form of demand function, see Arrow (1961) and Gorman (1963).

8. As is noted in the previous section, the series for demand often incorporate at least some portion of inventory changes in the using countries, whether for speculative or other purposes. To the extent that this is true, it is very difficult to represent adequately inventory behavior unless such inventory changes are highly correlated with the measurable ones.

9. The makeup and weighting of this index is discussed for each commodity in Chapter Three.

10. The alternative form of estimating relation (2-10) with different exponents for inventories and demand clearly implies other than constant returns to scale in the demand for inventories. If the absolute value of the exponent for demand is less than that for inventories, for example, increasing returns to scale are implied.

Chapter 3

Econometric Estimates of Commodity Models

The general specification which has been proposed for the models in Chapter Two has been used to construct eight agricultural commodity models for cocoa, coffee, tea, wool, cotton, sugar, wheat, and rice. This ordering obviously groups the commodities by their end uses—beverages,[1] textile-related, sugar, and basic grain staples. It also groups them roughly by the length of the gestation period required by the fixed inputs—tree and bush crops, animal products, and annual crops.

These commodities also differ greatly with regard to their origin and destination. Table 3-1 presents the percentage shares of world production and consumption for each region of the world in 1956 and 1971. The developing countries produced all of the world's output for two tropical crops—cocoa and coffee—more than half of world production for two others—tea and rice—and about half for sugar. Sugar is unique in that it is produced in the temperate areas as beet sugar and in the tropical zones as cane sugar. Consequently, it is produced in substantial proportions in all the regions. Cotton is also produced in all three areas, although it is not quite as climatically adaptable as sugar. Wool and wheat require a temperate climate and are produced primarily in the developed countries and in the colder areas of the centrally planned economies.

With regard to demand, in 1956 the developed countries were the dominant consumer group for all the commodities except for the basic staples of wheat and rice. Among the six nonstaples, they con-

Table 3-1. Production and Consumption of Commodities by Area[a]

		Cocoa	Coffee	Tea	Wool	Cotton	Sugar	Wheat	Rice
Share in World Production					(Percent of world production)				
Developed	1956	0	0	8	62	35	31	36	8
	1971	0	0	7	60	22	29	36	6
Developing	1956	100	100	74	21	36	50	19	54
	1971	100	100	72	19	43	48	20	56
Centrally Planned	1956	0	0	18	17	29	19	46	38
	1971	0	0	21	21	36	24	44	37
Share in World Consumption					(Percent of world consumption)				
Developed	1956	79	79	56	71	46	51	33	8
	1971	62	68	41	55	34	41	27	6
Developing	1956	16	21	31	9	23	30	22	54
	1971	23	29	41	15	30	34	27	56
Centrally Planned	1956	5	1	13	20	31	19	46	38
	1971	15	3	17	30	36	25	46	37

[a] Appendixes A and B give sources and data.

sumed less than half of the world total only in the case of cotton. For cocoa, coffee, and wool they absorbed over 70 percent of the total. Comparison with sources of production indicates that substantial net imports to the developed economies were involved for each of the six nonstaples. In the cases of the two staples, the centrally planned economies were most important in wheat consumption with 46 percent of the total (although the developed economies accounted for a third of world wheat demand) and the developing economies consumed 54 percent of the rice (with the centrally planned group accounting for 38 percent). Net movements among the three country groups were much less for the staples than for the other commodities, although not insignificant quantities of wheat flowed from the developed to the developing economies.

Over the sample period quite dramatic changes occurred in these relative demand shares—much more so than in the case of production. The shares of the developed countries in world demand dropped substantially for each of the eight commodities. For this group the 1971 shares as percentages of the 1956 shares ranged from 74 percent for cotton to 86 percent for coffee. The developed economies remained the dominant consuming group for coffee, cocoa, and wool—although in none of these cases did they any longer absorb as much as 70 percent of the world total. For tea both the developed and developing groups accounted for 41 percent each; for sugar the developed group consumed 41 percent; and for cotton the centrally planned economies absorbed 36 percent, as compared to 34 percent for the developed countries. The pattern for the two staples changed less substantially than for the other crops, although the developing countries increased their share significantly, primarily at the expense of the developed ones.

The shift in demand was to both developing and centrally planned economies. The former group increased its share for each of the eight commodities. The latter raised its share for all six of the nonstaples. The developing countries increased their net imports (or reduced their net exports) between 1956 and 1971 for each of the eight products except possibly for rice. Likewise the centrally planned economies increased their net imports (or reduced their net exports) for each of the commodities except for cotton and possibly rice. Thus, for both of these country groups these eight commodities became more of a net drain (or less of a source) of foreign exchange.

The remainder of this chapter considers the structure of the models estimated. There is one section for each commodity. Each follows the same pattern. Econometric estimates are presented and discussed first for supply relations for the developed mixed economies, the developing countries, and the centrally planned economies; then for demand relations for the same three groups; and finally for the price determination relation.[2]

COCOA[3]

Supply

The successful production of cocoa calls for very particular ecological conditions.[4] Consequently, the production of cocoa is concentrated in a relatively few countries within the developing group, primarily in Africa.

The supply equation for cocoa is:

$$ln\ PRO = 0.024\ ln\ PDF_{-6} + 0.080\ ln\ PDF_{-7} + 0.124\ ln\ PDF_{-8}$$
$$(0.5) \qquad\qquad (2.0) \qquad\qquad (3.2)$$

$$+ 0.112\ ln\ PDF_{-9} + 0.049\ T + 0.041\ DS7072 + 0.214\ DS65$$
$$(2.4) \qquad\qquad (7.6) \quad (1.6) \qquad\qquad (3.0)$$

$$+ 6.016$$
$$(41.4)$$

$$\bar{R}^2 = 0.94, SE = 0.061, DW = 2.2, 1956\text{--}1972 \tag{3-1}$$

This relation is consistent with a substantial part of the variance in the dependent variable over the 1956–1972 sample period, with no evidence of significant problems with serial correlation.

Although some earlier studies present evidence of significantly nonzero price response on the country level for the minor producers (e.g., Behrman [1968a]), there is no evidence of a significant short run price response on the aggregate level. On the other hand, there is evidence in our equation of a significantly nonzero long run response to the price ratio (*PDF*), with a mean lag of about eight years.[5] The start of this response after six or seven years is consistent with agronomic information about the time required for cocoa trees to reach their full maturity and with earlier econometric estimates (e.g. Bateman [1965a and 1965b] and Behrman [1968a]). The long run elasticity of 0.34 (with a *t*-value of 3.0), however, is smaller in value than the country estimates presented by Behrman

(1968a) for each of the five largest producers. This smaller responsiveness than estimated in the earlier studies probably reflects the extensions of barriers between international and domestic cocoa prices in the past decade.[6]

The significantly nonzero coefficient for the time trend indicates a steady secular growth in production at an annual exponential rate of 4.9 percent. This quite high growth rate probably reflects the considerable expansion of area devoted to cocoa production (especially in some of the French-speaking African nations), as well as the increase in disease control and prevention in older areas.[7]

The last variables included in the production relation are dummy variables for the special conditions in 1965 and in 1970–1972. Due in part to good weather, production in 1965 reached a level over 20 percent higher than in any previous years and not attained again until 1972. The significance of the coefficient of the dummy variable for 1965 and the size of the estimate reflect the importance of the special conditions in that year.

Demand

The consumption of cocoa is concentrated in countries with high per capita incomes. In 1971, for example, over 60 percent of the demand was in the developed mixed economy group, as compared to 23 percent in the developing countries and 15 percent in the centrally planned economies.[8] The ordering by growth rates over the 1955–1973 sample period, however, is just the reverse—so considerable diversification has occurred.[9]

The three demand functions are as follows:

developed economies

$$ln\,(D/POP) = -0.053\,ln\,PDF_{-0} -0.176\,ln\,PDF_{-1}$$
$$\quad\quad\quad\quad\quad (1.3) \quad\quad\quad\quad\quad (4.0)$$

$$-0.044\,ln\,PDF_{-2} -0.055\,ln\,PDF_{-3} + 0.192$$
$$\quad (1.0) \quad\quad\quad\quad (1.7) \quad\quad\quad\quad (21.1)$$

$$\bar{R}^2 = 0.93, SE = 0.033, DW = 1.1, 1955\text{–}1972 \quad\quad\quad (3\text{-}2)$$

developing economies

$$ln\,(D/POP) = -0.041\,ln\,PDF_{-1} + 0.427\,ln\,(GDP/POP)$$
$$\quad\quad\quad\quad\quad (1.0) \quad\quad\quad\quad\quad (1.3)$$

$$+ 0.678 \ln (D/POP)_{-1} -2.58$$
$$(3.4) \qquad\qquad (1.3)$$

$$\bar{R}^2 = 0.95, SE = 0.052, DW = 1.5, 1955\text{--}1972 \qquad\qquad (3\text{--}3)$$

centrally planned economies

$$\ln (D/POP) = -0.122 \ln PDF_{-0} -0.138 \ln PDF_{-1}$$
$$(2.9) \qquad\qquad (3.1)$$

$$-0.216 \ln PDF_{-2} -0.150 \ln PDF_{-3}$$
$$(5.1) \qquad\qquad (4.2)$$

$$+ 1.18 \ \ln (GDP/POP) -7.66$$
$$(27.4) \qquad\qquad (37.2)$$

$$\bar{R}^2 = 0.997, SE = 0.032, DW = 2.6, 1955\text{--}1972 \qquad\qquad (3\text{--}4)$$

On an overall level these relations are quite consistent with variations in the dependent variable, although the Durbin-Watson statistics suggest the possibility of some problems of serial correlation which may lead to inefficient estimates.

The major common determinant in these three relations is the distributed lag response to the deflated price.[10] The long run price elasticities are −0.33 (with an absolute value for the *t*-statistic of 11.1) for the developed economies, −0.13 for the developing economies, and −0.63 (with an absolute value for the *t*-statistic of 13.7) for the centrally planned economies. The mean lags for adjustment are 1.3, 3.1, and 1.6 years, respectively. The standard deviations are 0.65 and 0.24 years, respectively for the developed and centrally planned economies, so these two mean lags are not statistically different at the 5 percent level. The adjustment process is much slower, in contrast, in the developing countries, for which over three years are required before half of the adjustment occurs.

The relatively quick and quite substantial estimated price response of the centrally planned economies, prima facie, may be surprising because of their extensive use of non-price-allocative mechanisms. Similar results, however, are found for a number of commodities considered in this study. The explanation may lie in decision by authorities concerned less about buying certain quantities of a given good than about expending a certain amount on it, especially if foreign exchange is involved. If only the amount expended entered in, of course, the price elasticity of the quantity purchased would be

−1.0. Because a pure system of decisions to spend a certain amount does not prevail, the actual elasticity is smaller in absolute value—but still large relatively to those for the other country groups.

The relatively slow and small estimated price response of the developing economies for cocoa is also found for a number of other commodities. In this case the explanation probably lies in the widespread use of quantitative restrictions in the foreign sector regimes of developing nations.[11] Although cocoa is grown almost exclusively in developing countries, its production is very concentrated so that most developing countries must import most of what they use. Because foreign exchange shortages are pervasive, price-related policies are widely thought to be ineffective and since cocoa and cocoa products are considered not to be basic food necessities, quantitative restrictions frequently are applied. Any actual price responses on the part of the final users, therefore, may be poorly reflected in the estimates because the quantitative restrictions break the link between international and domestic prices and the former are used in the regressions.

The second important determinant of per capita demand emphasized in the previous chapter is the level of per capita income or per capita production. For the developed mixed economies the estimated relations reveal no evidence of a significantly nonzero response to income or production, although the implied elasticity of total demand with respect to population is one.[12] For the developing economies the estimates suggest slow adjustment toward an inelastic long run response—which is significantly nonzero only at the 10 percent level. For the centrally planned nations the estimates imply a clearly significant, elastic response. Given that this last group is midway between the other two in respect to the level of per capita product, these estimates provide some support (at least in the short run) for earlier hypotheses about the demand for cocoa having a flattened S shape when plotted against per capita product or income. At low levels cocoa is almost not considered among the set of consumption possibilities. At medium levels of income it is considered a luxury, with a high income elasticity. At high levels of income, satiation (on a per capita consumption basis) occurs, so, once again, there is no income response. Under such a pattern, future incremental markets beyond what is implied by population growth will be in the medium level per capita income countries and in the higher per capita income developing economies.

Price Determination

The estimated relationship for the deflated cocoa price[13] is exactly of the form discussed above:

$$\ln PDF = 0.858 \ln (STK/D) -0.026\ T -0.292$$
$$(9.2) \qquad\qquad (5.3) \qquad (2.3)$$

$$\bar{R}^2 = 0.87, SE = 0.116, DW = 1.5,\ 1955\text{-}1973 \qquad\qquad (3\text{-}5)$$

There is a downward secular trend of -2.6 percent per year in the deflated price, or, equivalently, of -3.0 percent per year in the proportionality factor between inventories and demand for a given deflated price. This probably reflects economies of scale as demand expands, together with the need for a smaller stock to demand ratio as communications and transportation improve.

The fluctuations in the deflated price around the secular trend are captured fairly well by the ratio of stocks to demand. The point estimate of -0.858 implies that the demand for inventories is quite responsive with respect to the deflated price with an elasticity slightly greater than unity ($-1.17 = -1/0.858$). The response also is rapid— basically to conditions in the current period.

COFFEE[14]

Supply

Coffee production, like that for cocoa, is concentrated in the developing nations because of ecological requirements. The principal traditional producing countries are in Latin America, particularly Brazil and Colombia, but African production has been expanding. The relationship for the developing nations as a whole is as follows:

$$\ln PRO = -0.023 \ln PDF_{-6} + 0.054 \ln PDF_{-7} + 0.142 \ln PDF_{-8}$$
$$(0.3) \qquad\qquad (1.0) \qquad\qquad (5.0)$$

$$+\ 0.153 \ln PDF_{-9} + 0.024\ T + 0.251\ DUM6566$$
$$(3.2) \qquad\qquad (3.9) \qquad (5.2)$$

$$+\ 0.241\ DUM60 + 10.470$$
$$(3.1) \qquad\qquad (69.5)$$

$$\bar{R}^2 = 0.87, SE = 0.068, DW = 2.8,\ 1956\text{-}1973 \qquad\qquad (3\text{-}6)$$

In most respects this relation is similar to that presented above for cocoa. It is consistent with a substantial portion of the variance over the 1956–1973 sample period. There is no evidence of a short run price response. There is evidence of a long run price response with a gestation period of six or seven years. The long run elasticity of 0.33 (with a *t*-value of 3.2) and the mean lag of 8.4 years, in fact, are not significantly different from the analogous estimates for cocoa.

The coefficient estimate for the time trend indicates an annual exponential secular growth rate of 2.4 percent for coffee. This is significantly lower than the estimate for cocoa. Of course, this point estimate combines the effects of much more rapid growth in the African producing areas with much slower expansion in most of the traditional Latin American suppliers.

Dummy variables for special conditions—primarily related to the weather—also are included for coffee production. These relate to the increase in 1960 of 27.9 percent to a level exceeded only once in the sample period and to the decline of 28.7 percent in 1965 followed by an increase of 61.3 percent in 1966 (or of 14.9 percent from the 1964 level). This latter experience seems clearly to relate to a shift between 1965 and 1966. In general, however, statistically significant evidence could not be found for the two year coffee cycle which is hypothesized in some earlier literature.[15]

Demand

The consumption of coffee is even more concentrated in countries with high per capita incomes than is the consumption of cocoa. In 1973, for example, developed mixed economies accounted for 66 percent of world coffee consumption, developing economies absorbed 30 percent, and the centrally planned economies used only 4 percent. In 1956 the respective percentages were 79, 21, and 1. Thus, as for cocoa, coffee demand has been gaining relatively most quickly in the centrally planned economies, with the developing economies next.[16] The three demand functions are as follows:

developed economies

$$ln\,(D/POP) = 3.264 + 0.197\,ln\,(GDP/POP) - 0.237\,ln\,PDF_{-1}$$
$$\quad\quad\quad (10.3)\quad (3.0)\quad\quad\quad\quad\quad (6.1)$$

$$\quad\quad - 0.198\,(ln\,PDF_{-1} - ln\,PDF_{-2})$$
$$\quad\quad\quad (3.9)$$

$$\bar{R}^2 = 0.94,\,SE = 0.030,\,DW = 2.0,\,1955\text{--}1972 \quad\quad\quad\quad (3\text{-}7)$$

developing economies

$$ln\,(D/POP) = 0.567 + 0.400\,ln\,(GDP/POP)_{-1} - 0.314\,[(ln\,PDF_{-1}$$
$$(0.5)\quad(1.8)\phantom{+0.400\,ln\,(GDP/POP)_{-1}}(3.4)$$

$$+\,ln\,PDF_{-2})/2.0]\,-0.242\,(ln\,PDF_{-2} - ln\,PDF_{-3})$$
$$\phantom{+\,ln\,PDF_{-2})/2.0]\,}(2.2)$$

$$\bar{R}^2 = 0.88,\,SE = 0.060,\,DW = 2.4,\,1955\text{-}1972 \qquad\qquad (3\text{-}8)$$

centrally planned economies

$$ln\,(D/POP) = -6.380 + 1.447\,ln\,(GDP/POP) - 1.247\,[(ln\,PDF_{-0}$$
$$(6.3)\quad(7.0)(5.3)$$

$$+\,ln\,PDF_{-1})/2.0]$$

$$\bar{R}^2 = 0.97,\,SE = 0.049,\,DW = 1.9,\,1955\text{-}1972 \qquad\qquad (3\text{-}9)$$

On an overall level these relations are quite consistent with variations in the dependent variables. The Durbin-Watson statistics do not suggest that serial correlation is a problem.

The first major common determinant in these relationships is the response to prices. The long run price elasticities are –0.24 for the developed economies, –0.31 for the developing economies, and –1.25 for the centrally planned economies. These estimates are substantially larger in absolute value than those reported above for cocoa in the case of the developing and centrally planned economies, although somewhat smaller for the mixed developed ones. Adjustment requires half a year for the centrally planned economies, a year for the mixed developed economies, and a year and a half for the developing economies.[17] This ordering is similar to that noted above for cocoa, although the adjustments in coffee demand generally are faster—especially for developing countries.

The relatively quick and quite substantial coffee price response of the centrally planned economies merits emphasis. This result again may reflect the kind of behavior discussed above in regard to cocoa. That the price elasticity for coffee is not significantly different from unity at the 10 percent level, in fact, suggests that the above explanation may be more relevant for coffee than for cocoa.

For the mixed developed and developing economies, in addition to the reaction to the deflated price level, there is evidence of a significant short run reaction to the logarithmic difference of lagged prices. This "accelerator" variable (in Harberger's [1963] terms) represents short run adjustments—perhaps in stocks—to price changes. Because of these responses, the short run price elasticities are larger in absolute value than the long run ones: –0.44 for the developed economies after one year and –0.55 for the developing economies after two years. Such demand behavior increases the probability of short run price stability, in the coffee market, *ceteris paribus.*

The second major determinant common to all three demand relations is the level of per capita product. The implied elasticities are 0.20 for the mixed developed economies, 0.40 for the developing economies, and 1.45 for the centrally planned economies. The response is significantly greater for coffee than for cocoa in the case of the developed countries, although there is no significant difference for the other two groups. The pattern across groups, even more than in the case of cocoa, suggests the existence of a flattened S shape for the income response of coffee demand. Finally, the adjustment to GDP changes is rapid in all three cases—but, once again, slowest in the developing economies.

Price Determination

The estimated relationship for the deflated coffee price[18] is exactly of the form discussed in Chapter Two:

$$\ln PDF = 0.362 + 0.097\ DUM6470 - 0.389\ \ln\ (STK/D) - 0.016\ T$$
$$\qquad\ \ (6.2)\quad (5.6)\qquad\qquad (11.2)\qquad\qquad (5.5)$$

$$\bar{R}^2 = 0.95, SE = 0.061, DW = 1.8, 1955\text{-}1972 \qquad\qquad (3\text{-}10)$$

There is a downward secular trend of –1.6 percent per year in the deflated price, or equivalently of –4.1 percent per year in the proportionality factor between inventories and demand for a given deflated price. As for cocoa, this steady decline probably reflects economies of scale in holding inventories and lessened relative inventory needs due to improved transportation and communications.

The fluctuations in the deflated price around the secular trend are captured quite well by the variance in the ratio of inventories to demand. The point estimate of –0.389 implies that the demand for coffee stocks is quite responsive with respect to the deflated price—the implied elasticity is –2.57 (= –1/0.389). For coffee much more than for cocoa, thus, fluctuations in supply and demand are absorbed in inventory changes relative to price movements.

The final factor which enters into the price determination relations is a dummy variable with monotonic declining values for 1964–1966 (i.e., 3, 2, 1) and a value of 2 for 1970. This variable represents the positive, but diminishing, effects of the international coffee agreement on prices in the mid-1960s and the possibility of renewal in 1970.

TEA[19]

Supply
Tea is more widely produced among the three regions than are the previous two beverage crops. Nevertheless, the developing economies, particularly India and Sri Lanka, were the dominant producing group during the sample period, although the centrally planned economies, largely China, gained somewhat at the expense of the other two groups. Between 1956 and 1971, for example, the developed group share fell from 8 to 7 percent, the developing countries' share fell from 74 to 72 percent, and the centrally planned economies' share rose from 18 to 21 percent. The supply functions for each of these groups are as follows:

developed economies

$$\ln PRO = 0.061 \ln PDF_{-4} + 0.447 \left[\ln (GDP/POP)_{-2} - \ln (GDP/POP)_{-3}\right]$$
$$(1.8) \qquad (1.4) \qquad (0.2) \qquad (0.7)$$

$$+ 0.018 \ T + 4.071$$
$$(6.0) \qquad (59.8)$$

$$\bar{R}^2 = 0.91, SE = 0.029, DW = 1.9, 1955\text{-}1972 \qquad (3\text{-}11)$$

developing economies

$$\ln PRO = 0.146 \ln PDF_{-1} + 0.075 \ln PDF_{-6} + 0.083 \ln PDF_{-7} + 0.053 \ln PDF_{-8}$$
$$(1.0) \qquad (1.7) \qquad (1.8) \qquad (1.3)$$

$$+ 0.015 \ln PDF_{-9} + 0.040 \ T + 5.939$$
$$(0.4) \qquad (5.6) \qquad (43.9)$$

$$\bar{R}^2 = 0.95, SE = 0.037, DW = 2.4, 1956\text{-}1972 \qquad (3\text{-}12)$$

centrally planned economies

$$\ln PRO = 0.234 \ln PDF_{-4} + 0.261 \ln PDF_{-5} + 0.171 \ln PDF_{-6} + 0.054 \ln PDF_{-7}$$
$$(3.7) \qquad (4.1) \qquad (2.7) \qquad (0.9)$$

+ 0.056 T + 4.223
  (10.1)   (33.2)

$\bar{R}^2 = 0.95, SE = 0.053, DW = 1.6, 1955–1972$ (3-13)

For all three groups the estimated relations are consistent with quite substantial proportions of the variance in the dependent variables, with no evidence of substantial problems with serial correlation.

The extent of price response varies considerably across the three groups. For the developing countries there is weak evidence of a short run response with an elasticity of 0.15, but the standard error of this point estimate is almost as large. For the other two groups there is not even such weak evidence for a short run price response.

The long run price elasticities are 0.089 (with a t-value of 0.9) for the developed group, 0.23 (with a t-value of 1.7) for the developing countries, and 0.72 (with a t-value of 3.7) for the centrally planned economies. As in the cases of demand for cocoa and coffee, thus, these estimates imply that the centrally planned economies have more substantial and more significant price responses than do the other two groups. The explanation may be related to that posited for demand. In order to save scarce foreign exchange, domestic production of desired commodities with high international prices is encouraged by inducements or required by quantitative measures. The effective result, however, is equivalent to a positive supply response to world market price conditions.

The mean lags for the long run adjustments to prices are 4, 7, and 5.1 years, respectively. They imply a more rapid supply response to changes in market conditions for tea than for the two tree crops discussed above—although not as rapid as for most of the products considered below.

For all three groups both the short run and the long run price responses occur around significant positive secular trends—1.8 percent per year in the developed countries, 4.0 percent per year in the developing countries, and 5.6 percent per year in the centrally planned economies. For the developing countries this rate is greater than that for coffee, but slightly less than that for cocoa. The relatively high secular growth rates for tea production in the developing and centrally planned economies reflects the steady increase of inputs used for these crops in these two area.

For tea, in contrast to cocoa and coffee, there is no evidence of statistically significant response to special weather conditions.

Demand

The absorption of tea has been less concentrated in the developed economies than either the consumption of cocoa or that of coffee. In 1956 consumption shares were 56 percent for the developed group, 31 percent for the developing countries, and 13 percent for the centrally planned economies. The last two groups increased their demand relatively quickly, so that for 1971 the respective shares were 41, 41, and 17 percent. By the end of the period, thus, the developing economies were as important in tea absorption as the developed countries.

The demand functions estimated for these three groups are as follows:

developed economies

$$\ln (D/POP) = -0.072 \ln PDF_{-0} + 0.447 (\ln (GDP/POP)_{-2} - \ln (GDP/POP)_{-3})$$
$$\quad\quad\quad (1.3) \quad\quad\quad\quad\quad (2.0)$$

$$- 0.005\ T - 0.313$$
$$\quad (1.9) \quad (6.9)$$

$$\bar{R}^2 = 0.38, SE = 0.014, DW = 1.9, 1955\text{-}1971 \quad\quad\quad (3\text{-}14)$$

developing economies

$$\ln (D/POP) = -0.140 \ln PDF_{-2} - 0.201 (\ln PDF_{-2} - \ln PDF_{-3})$$
$$\quad\quad\quad (1.4) \quad\quad\quad\quad (2.5)$$

$$+ 0.431 (\ln GDP/POP) - 3.393$$
$$\quad (2.4) \quad\quad\quad\quad (4.1)$$

$$\bar{R}^2 = 0.90, SE = 0.035, DW = 2.2, 1955\text{-}1971 \quad\quad\quad (3\text{-}15)$$

centrally planned economies

$$\ln (D/POP) = -0.277 \ln PDF_{-1} + 0.382 \ln (D/POP)_{-1} - 1.179$$
$$\quad\quad\quad (2.7) \quad\quad\quad\quad (3.4) \quad\quad\quad\quad (5.2)$$

$$\bar{R}^2 = 0.88, SE = 0.060, DW = 1.8, 1955\text{-}1971 \quad\quad\quad (3\text{-}16)$$

These relations—especially that for the developed group—provide less adequate explanation of the variations of the dependent variables than those for cocoa and coffee. In the case of the developed countries, the low coefficient of determination probably reflects that the series for the dependent variable appears to be generated by random disturbances around a very slight secular downward trend. In none of the three cases is serial correlation obviously a problem. However, for the centrally planned economies the Durbin-Watson statistic is biased towards 2.0 by the inclusion of a lagged dependent variable, so it provides little information.

The first major demand determinant is the response to prices. The long run price elasticities are −0.07 for the developed economies, −0.14 for the developing countries, and −0.48 for the centrally planned countries. The first two of these are significantly nonzero only at the 10 percent level. In addition there is a significantly nonzero expectations or accelerator response for the developing economies which implies a short run price elasticity as large as −0.34 (after two years).

The relative pattern of long run price responses is similar to that for coffee, once again with the greatest response in the centrally planned economies. However the magnitudes of the elasticities are significantly smaller than the above presented estimates for either coffee or cocoa (with the single exception of the developing countries in the case of cocoa). *Ceteris paribus* this result implies greater price variation for tea than for the other two products.

The second major determinant is the level of per capita GDP. In sharp contrast to the "S curve" estimates discussed above for cocoa and coffee, only for the developing economies is there evidence of a significantly nonzero long run response to this factor. For the developing countries the estimated long run income elasticity is 0.43 (which is not significantly different from the cocoa and coffee income elasticities for this group). For the developed mixed economies, in addition, there is a significantly nonzero short run response to changes in per capita product of the same order of magnitude.

The third demand determinant is a downward secular trend of −0.5 percent per year for the developed countries. This decline presumably reflects a shift in aggregate tastes against tea.

Price Determination

The estimation of a satisfactory price[20] determination relation for tea presents certain difficulties. Two examples of the results obtained by using the specification of relation (2-10) follow. In both cases a geometric distributed lag adjustment is assumed, but the price is deflated in the first and undeflated in the second. In neither case is a trend included.

$$\ln PDF = -1.128 \ln (STKW/DW) + 0.907 \ln PDF_{-1} + 1.463$$
$$\quad\quad\quad (3.3) \quad\quad\quad\quad\quad\quad (20.3) \quad\quad\quad (3.2)$$

$$\bar{R}^2 = 0.98, SE = 0.040, DW = 1.9, 1955-1971 \quad\quad\quad\quad (3-17)$$

$$\ln P = -0.783 \ln (STKW/DW) + 0.791 \ln P_{-1} + 1.996$$
$$\quad\quad\quad (2.6) \quad\quad\quad\quad\quad\quad (9.0) \quad\quad\quad (3.1)$$

$$\bar{R}^2 = 0.88, SE = 0.038, DW = 1.9, 1955-1971 \quad\quad\quad\quad (3-18)$$

For both relations the coefficients of determinations suggest reasonable consistency with variations in the dependent variable over the sample period. In both cases, of course, the Durbin-Watson statistic is biased toward two because of the inclusion of lagged dependent variables.

The implied sensitivity of the deflated and undeflated tea price to the level of inventories relative to demand in the two cases is quite considerable. The respective short run elasticities are −1.1 and −0.8. The respective long run elasticities are −12.1 and −3.7! The converse of these large long run elasticities, of course, is that the long run response of tea inventories with respect to tea prices is quite small—the respective elasticities are −0.1 and −0.3.

These results, together with those presented above for tea supply and demand, prima facie seem to imply that the tea price is affected much more by a random change of a given percentage in supply or demand than are those for cocoa and coffee. The aggregate long run price elasticities for supply for these three crops are about the same, but the long run price elasticities for both current and inventory demand are much smaller for tea than for the other two. Therefore it would seem that prices must absorb much more of a random shock in the tea market than in the others. However the smaller absolute sizes of the demand elasticities in the tea market are offset within the simulated market system by the relatively quick price response of consumption demand and, more importantly, the relatively slow adjustment in the price relations. Therefore the simulations discussed below suffer, if anything, from too little immediate price response.

As is suggested at the start of this discussion of price determination in the tea market, these estimates present several problems.

1. The adjustment period is very long for both relations. The respective mean lags, to be more explicit, are 9.6 and 3.8 years.
2. Both relations seem systematically to underestimate tea prices for 1964–1968 on a single equation basis.
3. Within a general equilibrium framework neither relation tracks as well as might be desired. Once again there is systematic underestimation of the tea price in the middle and late 1960s.
4. The a priori rationale for the undeflated price in relation (3–18) is not clear. If the price deflator is added as another right-hand side variable, moreover, its collinearity with the other variables results in an a priori implausible positive coefficient for the logarithm of the inventory to demand ratio.

For these reasons, considerable efforts were expended in exploring alternative price determination functions. Secular trends, different lag distributions, accelerator variables, the pound sterling to the United States dollar exchange rate (since much of tea trade is within the sterling area), and differing coefficients for stocks and for demand are a few examples of the alternatives considered. On both a single equation and complete system basis, none of these dominate relations (3–17) and (3–18)—and most are poorer. Relation (3–17) is used in the multiplier simulations below.

WOOL

A number of econometric studies have investigated the dynamics of the world wool market.[21] These studies have emphasized the substitution of synthetic fibers for wool as inputs into textile mill consumption on the demand side and have recognized the relation between wool shearing, the stock of sheep, and the raising of lambs on the supply side.

Supply
Wool is a product largely of the temperate zones, with approximately 44 percent of production of the developed market economies originating in Australia and New Zealand and 7 percent of the production of the developing market economies concentrated in Argentina.[22] Wool production shows little short run adjustment to price, as the shear depends greatly on the size of the existing population of sheep. The adjustment of the sheep population to expected price is

a phenomenon which takes some years and which may be indistinguishable in some cases from the time trend. The wool supply equations are as follows:

developed economies

$$\ln PRO = 0.014\, T + 0.039\, \ln PDF_{-0} + 0.102\, (\ln PDF_{-0} - \ln PDF_{-1})$$
$$\ (2.7)\quad\ \ (0.4)\qquad\qquad\quad (1.1)$$

$$+\ 6.577$$
$$\ (86.9)$$

$$\bar{R}^2 = 0.75, SE = 0.038, DW = 1.5, 1955\text{-}1971 \qquad\qquad (3\text{-}19)$$

developing economies

$$\ln PRO = 0.043\, T + 0.026\, \ln PDF_{-2} + 0.122\, \ln PDF_{-3} + 0.204\, \ln PDF_{-4}$$
$$\ (7.3)\quad\ \ (0.5)\qquad\qquad (2.7)\qquad\qquad (5.7)$$

$$+\ 0.191\, \ln PDF_{-5} + 4.933$$
$$\ (4.5)\qquad\qquad (47.9)$$

$$\bar{R}^2 = 0.90, SE = 0.034, DW = 1.8, 1955\text{-}1971 \qquad\qquad (3\text{-}20)$$

centrally planned economies

$$\ln PRO = 0.013\, T + 0.739\, \ln PRO_{-1} + 0.010\, \ln PDF_{-1} + 0.037\, \ln PDF_{-2}$$
$$\ (2.0)\quad\ \ (4.8)\qquad\qquad (0.2)\qquad\qquad (0.9)$$

$$+\ 0.059\, \ln PDF_{-3} + 0.054\, \ln PDF_{-4} + 1.294$$
$$\ (1.4)\qquad\qquad (1.2)\qquad\qquad (1.6)$$

$$\bar{R}^2 = 0.97, SE = 0.032, DW = 2.2, 1955\text{-}1971 \qquad\qquad (3\text{-}21)$$

In the developed mixed economies, the short run price elasticity is approximately 0.14 but the long run elasticity tends toward 0.04. Such a low response is understandable only in the light of an upward time trend of 1.4 percent per year, which offsets the effect of the substantial decline of the real wool price which occurred over the sample period. For the developing mixed economies, including Argentina, somewhat greater price responses and longer lags are apparent. While there is no evidence of a short run price response in the developing economies, the elasticity builds up to 0.54 over a five year lag period. For the centrally planned economies the equation combines a distributed lag and dependent variable to yield a long run elasticity of 0.61, again with a long period of lag response.

Demand

This study did not attempt to consider the direct relations between textile production and the derived demand for inputs of wool. Such a process would involve complex inventory adjustment procedures that would call for inventory statistics that were not readily available. We do recognize the substitution of synthetic fibers for wool as a consequence of technical developments in the qualities of these fibers and the price of wool relative to its substitutes. The demand equations here estimated for wool are as follows:

developed economies

$$\ln (D/POP) = 1.807 \ln (GDP/POP) + 0.546 \ln (STKW/POPW)_{-1} - 0.064\, T$$
$$(3.2) \qquad\qquad (5.6) \qquad\qquad\qquad (3.4)$$

$$- 0.032 \ln (PW/PSYN)_{-2} - 0.070 \ln (PW/PSYN)_{-3}$$
$$(0.5) \qquad\qquad\qquad (1.2)$$

$$- 0.092 \ln (PW/PSYN)_{-4} - 0.076 \ln (PW/PSYN)_{-5} - 5.954$$
$$(1.7) \qquad\qquad\qquad (1.4) \qquad\qquad\qquad (2.7)$$

$$\bar{R}^2 = .87, SE = 0.031, DW = 1.0,\ 1955\text{–}1971 \qquad\qquad (3\text{-}22)$$

developing economies

$$\ln (D/POP) = 1.052 \ln (GDP/POP) - 0.195 \ln (PW/PSYN)_{-0}$$
$$(4.7) \qquad\qquad\qquad (1.6)$$

$$- 0.131 \ln (PS/PSYN)_{-1} - 0.031 \ln (PW/PSYN)_{-2}$$
$$(1.3) \qquad\qquad\qquad (0.3)$$

$$+ 0.129 \ln (PW/PSYN)_{-3} - 7.108$$
$$(0.9) \qquad\qquad\qquad (7.2)$$

$$\bar{R}^2 = 0.81, SE = 0.082, DW = 1.2,\ 1955\text{–}1971 \qquad\qquad (3\text{-}23)$$

centrally planned economies

$$\ln (D/POP) = 2.046 \ln (GDP/POP) - 0.096\, T + 0.002 \ln (PW/PSYN)_{-3}$$
$$(4.3) \qquad\qquad\qquad (3.6) \qquad (0.0)$$

$$- 0.040 \ln (PW/PSYN)_{-4} - 0.084 \ln (PW/PSYN)_{-5}$$
$$(0.9) \qquad\qquad\qquad (2.4)$$

$$- 0.085 \ln (PW/PSYN)_{-6} - 8.918$$
$$(2.0) \qquad\qquad\qquad (5.1)$$

$$\bar{R}^2 = 0.96, SE = 0.028, DW = 2.3,\ 1956\text{–}1971 \qquad\qquad (3\text{-}24)$$

It is apparent that wool consumption increases more than proportionately with per capita income. Wool represents a "luxury" product. In fact, in the developed mixed economies and in the centrally planned economies the elasticity with respect to income is 1.8 and 2.0 respectively. In the developing economies, which are principally tropical areas, the elasticity is considerably lower at 1.05. The substitution of other fibers for wool is captured by a long distributed lag on the relative price of wool and synthetic fibers. The substitution elasticity adds up to –0.27 distributed over a lag period of two to five years in the developed mixed economies and –0.20 over a lag of from three to six years in the centrally planned economies. In the developing economies, where wool is considerably less important, the long run substitution response is approximately –0.16. It is reasonable to assume that technological innovations in the characteristics and applications of substitute synthetics account for the negative time trends in the demand for wool amounting to 6 percent per year in the developed mixed economies and almost 10 percent per year in the centrally planned economies.

Inadequacies in the data suggested the introduction of a stock variable, *STKW/POPW*, in the demand equations for the developed economies. The series for wool consumption was computed by adding net imports to domestic production and subtracting estimated stock change; it is in effect a series for "domestic disappearance." When wool inventory levels are high, the stock change may be small or even negative. In that case, given a fairly even flow of imports, the calculated demand statistic is higher. The relation between stock levels and consumption may be simply a result of our calculation procedure.

Price Determination

The price[23] equation for wool separates the impact of stocks and of demand.

$$\ln PDF = -0.090\,T - 0.660\,\ln STKW + 1.697\,\ln DW + 0.275\,DUM$$
$$\quad\;\;(7.9)\qquad(2.8)\qquad\qquad(3.0)\qquad\qquad(4.3)\qquad\;\;6364$$

$$-6.996$$
$$(1.9)$$

$$\bar{R}^2 = 0.95,\, SE = 0.084,\, DW = 0.6,\, 1955\text{--}1971 \qquad\qquad (3\text{--}25)$$

Seen as a demand for inventories equation, as equation (2–11) above, there is a demand elasticity for inventories of -1.5 and an elasticity with respect to world demand of 1.6. There is, however, a systematic downward adjustment over time in price. For the years 1963 and 1964, which represent a period of attempted price stabilization, a dummy variable with a positive coefficient of 0.28 is obtained.

COTTON

The world cotton economy is greatly influenced by United States agricultural policy, since United States production dominates output from the developed market economies and contributes greatly to world supply. It is not clear to what extent U.S. agricultural policy has influenced the fluctuations of cotton price.

There have been few econometric studies of cotton.[24]

Supply

The supply of cotton in the developed market economies shows some downward movement during the sample period. On the other hand, supply in the less developed market economies shows a very sharp upward trend. The supply equations are as follows:

developed economies

$$\ln PRO = 0.063 \, T + 1.350 \, \ln (PUSCOT/USPDF)_{-1} + 1.061 \, \ln USALLOC$$
$$\quad\quad (2.9) \quad\quad (4.1) \quad\quad\quad\quad\quad\quad\quad\quad\quad\quad (1.5)$$

$$+ \, 5.571$$
$$\quad (2.7)$$

$$\bar{R}^2 = 0.64, SE = 0.125, DW = 1.64, \, 1955\text{-}1971 \quad\quad\quad\quad (3\text{-}26)$$

developing economies

$$\ln PRO = 0.037 \, T + 0.070 \, \ln PDF_{-1} - 0.146 \, DUM \, 71 + 7.723$$
$$\quad\quad (8.7) \quad\quad (0.7) \quad\quad\quad (3.1) \quad\quad\quad (94.2)$$

$$\bar{R}^2 = 0.96, SE = 0.043, DW = 1.6, \, 1955\text{-}1973 \quad\quad\quad\quad (3\text{-}27)$$

centrally planned economies

$$\ln PRO = 0.033 \, T + 0.571 \, \ln PRO_{-1} + 0.437 \, \ln PDF_{-3} + 2.829$$
$$\quad\quad (2.5) \quad\quad (3.3) \quad\quad\quad\quad (2.1) \quad\quad\quad\quad (2.2)$$

$$\bar{R}^2 = 0.82, SE = 0.081, DW = 1.3, \, 1955\text{-}1973 \quad\quad\quad\quad (3\text{-}28)$$

In view of the dominance of the United States as a producer among the developed mixed economies, the relevant equation incorporates variables specific to the United States. The price variable is the United States price of cotton relative to the United States GNP deflator, with a lag of one year. Thus, U.S. cotton farmers take into consideration last year's price in making their production decisions with an elasticity of 1.35. They have, however, also been constrained by a system of acreage allocations, a variable which enters explicitly in the equations with an elasticity near unity. In the developing economies, the upward time trend is dominant and the price response is very small, amounting to an elasticity of only 0.07. A dummy variable for 1971 accounts for unfavorable weather conditions. The equation for the centrally planned economies shows a relatively powerful price effect, a long run elasticity of 0.77 but with a substantial lag. The initial price response lags three years and the long run effect takes considerably longer. It is noteworthy, however, that cotton production in developing economies and in the centrally planned economies is principally from irrigated fields which require a long gestation period and a considerable investment to develop, whereas in the United States cotton is produced largely without irrigation and is readily substitutable for other crops.

Demand

As with other fibers, in the case of cotton we expect to find significant substitution with the new synthetic substitutes. The demand equations for cotton are as follows:

developed economies

$$ln (D/POP) = 0.475 \ ln \ (D/POP)_{-1} - 0.027 \ T - 0.230 \ ln \ (PCOT/PSYN)_{-1}$$
$$(2.7) \qquad\qquad\qquad (2.2) \qquad (2.9)$$

$$+ \ 0.603 \ ln \ (GDP/POP) - 1.365$$
$$(1.8) \qquad\qquad\qquad (1.0)$$

$$\bar{R}^2 = 0.92, SE = 0.025, DW = 1.6, 1955\text{-}1973 \qquad\qquad (3\text{-}29)$$

developing economies

$$ln \ (D/POP) = 0.471 \ ln \ (GDP/POP) - 0.021 \ ln \ (PCOT/PSYN)_{-1}$$
$$(15.0) \qquad\qquad\qquad (0.9)$$

$$- \ 0.046 \ ln \ (PCOT/PSYN)_{-2} - 0.060 \ ln \ (PCOT/PSYN)_{-3}$$
$$(2.9) \qquad\qquad\qquad (4.2)$$

$$- 0.050 \ln (PCOT/PSYN)_{-4} - 1.564$$
$$\quad (2.2) \qquad\qquad\qquad\quad (10.7)$$

$$\bar{R}^2 = 0.98, SE = 0.013, DW = 2.7, 1955\text{-}1973 \qquad\qquad (3\text{-}30)$$

centrally planned economies

$$\ln (D/POP) = 0.003 \, T - 0.108 \ln (PCOT/PSYN)_{-1} + 0.604 \ln (PRO/POP)$$
$$\quad\quad\quad (3.2) \qquad (2.6) \qquad\qquad\qquad\qquad (12.4)$$

$$+ 0.197 \ln (D/POP)_{-1} + 0.237$$
$$\quad (2.9) \qquad\qquad\qquad (4.0)$$

$$\bar{R}^2 = 0.97, SE = 0.019, DW = 1.9, 1955\text{-}1973 \qquad\qquad (3\text{-}31)$$

In the developed market economies, we note a geometric distributed lag on *GDP/POP* and on the price of cotton relative to the price of synthetics. The elasticity with respect to *GDP* is 0.6 in the short run and approximately 1.1 in the long run. The substitution elasticity for price is only –0.2 after a one year lag but builds up to a long term elasticity of –0.44. There is a negative time trend which probably reflects nonprice aspects of the substitution of synthetic fibers for cotton.

In the developing market economies, the elasticity with respect to GDP is of the same general order, approximately 0.5. The distributed lag on relative price is considerably longer, building up to a relatively low elasticity of 0.18 after four years. The potentials for substituting synthetic fibers are considerably smaller in the less developed than in the developed economies.

The approach for the centrally planned economies recognizes the tendency for these countries to rely largely on their own resources. Consumption of cotton is greatly dependent on domestic production with imports making up only part of the shortfall when domestic harvests are reduced. Thus a 1 percent change in the domestic harvest will affect consumption by 0.6 percent. There is evidence of price-induced substitution by synthetics, but the effect is small and operates with a lag largely of only one year. (There is a small distributed lag effect.) It is interesting to note that the time trend in centrally planned economies is positive. Consumption grows over time suggesting again that there is not the same tendency for technological substitution of synthetics as in the developed market economies.

Price Determination

The price[25] equation for cotton poses some difficult problems:

$$\ln PDF = 0.917 \ln PDF_{-1} - 0.043 \ln (STKW/DW)_{-0} - 0.119 \ln (STKW/DW)_{-1}$$
$$\quad (11.8) \qquad\qquad (0.5) \qquad\qquad\qquad (2.2)$$

$$\quad - 0.173 \ln (STKW/DW)_{-2} - 0.152 \ln (STKW/DW)_{-3} - 0.367$$
$$\qquad (3.2) \qquad\qquad\qquad (2.0) \qquad\qquad\qquad (3.7)$$

$$\bar{R}^2 = 0.93, SE = 0.06, DW = 1.5, 1953\text{-}1972 \qquad\qquad (3\text{-}32)$$

While the response of price to the stock-demand ratio is in the right direction, the estimated impact is very slow. Not only is the principal impact apparent with lags of one to three years but the lag dependent variable carries a coefficient of 0.9. This means that the price in the current year is largely influenced by the lagged price and that current events have almost no influence on price. Moreover the lag effect is extended very far backward in time. This may not be a realistic result and may reflect the systematic downward trend of the *PDF* variable over the sample period. As we see below, this slow adjustment causes difficulties in the simulation performance of the model, particularly in the multiplier simulations.

We should also note that the price used in the supply equations for the developed market economies, the U.S. price, is not linked to the world price. This effectively isolates production in the developed market economies from the rest of the model, but probably is not unrealistic in the light of U.S. agricultural policy over the sample period.

SUGAR

Perhaps because of the complexity of the institutional restrictions in this market,[26] there have been few serious attempts at econometric modeling of the sugar market. Tewes (1972) produced a small forecasting model which includes equations for price, consumption, and acreage. Wymer (1975) provides an interesting attempt to fit a continuous differential equation model, but leaves production as exogenous. Other models have focused on spatial equilibrium analysis rather than production or simulation over time.

Supply

Sugar is unique among agricultural products in being a crop both of the tropics (sugar cane) and of the colder parts of the temperate areas (sugar beets). The major part of the production from the developed countries and from the centrally planned economies is from beets, although Cuba is an important cane sugar producer. The bulk of the crop in the developing countries is cane sugar.

The production equations for sugar are as follows:

developed economies

$\ln PRO = 0.042\ T + 0.152\ \ln PDF_{-1}$
$\qquad\quad (12.1)\qquad (1.6)$

$\qquad -0.111\ (\ln PDF_{-1} - PDF_{-2}) + 1.434$
$\qquad\ (1.3)\qquad\qquad\qquad\qquad (3.3)$

$\qquad \bar{R}^2 = 0.95, SE = 0.055, DW = 2.3,\ 1955\text{-}1973 \qquad\qquad (3\text{-}33)$

developing economies

$\ln PRO = 0.039\ T + 0.193\ \ln PDF_{-2}$
$\qquad\quad (16.9)\qquad (3.3)$

$\qquad -0.087\ (\ln PDF_{-2} - \ln PDF_{-3}) + 0.091\ DUM\ 3$
$\qquad\ (1.8)\qquad\qquad\qquad\qquad\quad (5.7)$

$\qquad +1.78$
$\qquad\ (6.4)$

$\qquad \bar{R}^2 = 0.98, SE = 0.032, DW = 1.1,\ 1955\text{-}1973 \qquad\qquad (3\text{-}34)$

centrally planned economies

$\ln PRO = 0.063\ T + 0.262\ \ln PDF_{-2}$
$\qquad\quad (4.6)\qquad (1.5)$

$\qquad -0.021\ (\ln PDF_{-2} - \ln PDF_{-3}) + 0.451\ \ln PRO_{-3}$
$\qquad\ (0.1)\qquad\qquad\qquad\qquad\quad (2.6)$

$\qquad -0.126$
$\qquad\ (0.1)$

$\qquad \bar{R}^2 = 0.94, SE = 0.084, DW = 1.2,\ 1955\text{-}1973 \qquad\qquad (3\text{-}35)$

The fit of these equations is reasonably good.

It is notable that the supply responses in the developing countries are slower than in the developed economies.[27] This is in line with the production period of some 18–24 months for new sugar cane acreage. The supply response in the centrally planned economies must be seen as an adjustment of supply planning to world market considerations.

The supply elasticity responses and their time lags are as follows:

	Time Lag			
	Effect of Value with Lag of:			Cumulative Long Run Effect
	One Year	Two Years	Three Years	
Developed economies	0.041	0.111	0	0.152
Developing economies		0.106	0.087	0.193
Centrally planned economies		0.241	0.472	0.713

The impact of crop variations in the developing economies in 1961 and 1963 pose a difficult problem for production estimates. In this instance, we have introduced a dummy variable to account for weather-related crop fluctuations, but considerable fluctuations remain that can only be associated with the vagaries of weather. It would be a difficult project, outside the objectives of this study, to pursue the relationship between sugar output and weather characteristics such as temperature and rainfall. Accounting in detail for weather factors would not help for prediction since one would have to have the relevant weather forecasts. However, we do investigate the effect of random error in supply corresponding to weather variations in the model simulations below.

Demand
The demand equations are as follows:

developed economies

$$ln\,(D/POP) = 0.080\ ln\,PDF + 0.184\ ln\,(GDP/POP)$$
$$\quad\quad (1.4) \quad\quad\quad\quad\quad (3.0)$$

$$\quad\quad + 0.401\ ln\,(D/POP)_{-1} + 1.400$$
$$\quad\quad\quad (2.2) \quad\quad\quad\quad\quad (3.5)$$

$$\bar{R}^2 = 0.97,\, SE = 0.012,\, DW = 2.1,\, 1955\text{-}1973 \quad\quad\quad\quad (3\text{-}36)$$

developing economies

$$ln\,(D/POP) = -0.048\ ln\,PDF_{-1} + 0.779\ ln\,(GDP/POP)$$
$$\quad\quad (2.1) \quad\quad\quad\quad\quad (21.2)$$

$$\quad\quad -0.965$$
$$\quad\quad (4.0)$$

$$\bar{R}^2 = 0.98,\, SE = 0.018,\, DW = 2.4,\, 1955\text{-}1973 \quad\quad\quad\quad (3\text{-}37)$$

centrally planned economies

$$ln\ (D/POP) = -0.119\ ln\ PDF + 0.089\ ln\ (GDP/POP)$$
$$ (2.3) (0.9)$$

$$+\ 0.747\ ln\ (D/POP)_{-1} + 0.749$$
$$ (6.0) \phantom{\ ln\ (D/POP)_{-1} +\ } (2.2)$$

$$\bar{R}^2 = 0.97, SE = 0.045, DW = 1.3,\ 1955\text{--}1973 \tag{3-38}$$

These functions give an excellent explanation for the movement of sugar demand per capita. The demand elasticities fall into reasonable patterns as follows:

	Price Elasticities		Income Elasticities	
	Short Term	Long Term	Short Term	Long Term
Developed economies	−0.02	−0.04	−0.18	−0.30
Developing economies	−0.05	−0.05	−0.78	−0.78
Centrally planned economies	−0.12	−0.48	−0.09	−0.36

The price elasticities are very small everywhere. This is in line with our expectations. Sugar is an essential commodity used primarily as an input (or as a compliment to) other products. Indeed, it is notable to see significant or near significant price responses since domestic sugar prices frequently do not move in correspondence with world prices. The elasticities with respect to income also are very small in the developed and centrally planned economies. But they are considerably larger, though still below unity, in the developing economies where the standard of living is lower and where sugar use is still considerably below saturation levels.[28]

Price Determination

The United Nations export price index covers approximately 75 percent of sugar in international trade.[29] However, it provides only an inadequate measure of domestic sugar prices and does not directly include beet sugar prices at all. The latter, however, can be expected to move with the price of cane sugar.

The price determination mechanism takes into account the relationships between available inventory stocks and the level of demand. In the case of sugar this has been formulated as a nonlinear relationship:

$PDF = 43.710 + 3.627 \ [1.0/(STK/DW) - Min \ (STK/DW) + 0.1]$
    (4.5)   (2.1)

$-2.788 \ (STK - STK_{-1}) - 11.892 \ DUMDS_{-1}$
  (2.5)   (-1.7)

$\bar{R}^2 = 0.68, SE = 9.806, DW = 2.1, 1955\text{-}1973$ (3-39)

Alternative formulations of this equation were considered. The one selected here contains a non-linearity on the stock-demand ratio in order to recognize the heightened price response when the ratio of stock to demand is near its historical lows. This non-linearity was introduced in an attempt to catch the sudden upsurges of sugar prices. The non-linear effect works in the equation to give an impact of 36 index points on price for every .01 change in the stock/demand ratio (STK/DW) when the latter is near its historic low of .15(Min (STK/DW)). When STK/DW is at .45, a very high level, the impact on price is only 9 index points. However, this formulation does not yield an equation which is significantly better than one assuming a linear relationship between stock/demand ratio and price. Disappointingly, it does not suffice to explain the 1973 price upsurge, though it goes a substantial direction toward explaining the buildup of prices at that time. The change in stock variable is an attempt to catch involuntary inventory accumulation. Thus if inventories increase the effect is to reduce price given the stock/demand ratio. The dummy variable explains a sharp increase of prices in 1957, 1964 and 1973.

WHEAT[30]

Supply
Substantial quantities of wheat are produced in all three groups of countries. Over the sample period the relative shares did not change much, although the developing countries did increase their proportion somewhat at the expense of the centrally planned economies due to the expansion of the Mexican dwarf varieties in the former. To be more explicit, between 1956 and 1971 the developing countries' share increased from 19 to 20 percent, the centrally planned group share fell from 46 to 44 percent, and the developed economies share stayed almost constant at 36 percent. Note that in contrast to the products discussed above, the centrally planned economies have the largest share of production. In the international markets, however,

the largest exporters are the developed countries and Argentina in the developing group.

The supply functions for each of the country groups are as follows:

developed economies

$$\ln PRO = 0.340 \ln PDF_{-0} + 0.152 \ln PDF_{-1} - 0.112 \ln PDF_{-2} + 0.040\ T$$
$$(1.8)(1.4)(-0.7)(4.5)$$

$$- 0.142\ DUM\ 6170 + 4.005$$
$$(3.8)(27.0)$$

$$\bar{R}^2 = 0.93,\ SE = 0.044,\ DW = 2.2,\ 1955\text{-}1971 \tag{3-40}$$

developing economies

$$\ln PRO = 0.117 \ln PDF_{-3} + 0.173 \ln PDF_{-4} + 0.170 \ln PDF_{-5} + 0.111 \ln PDF_{-6}$$
$$(0.7)(1.6)(1.6)(0.7)$$

$$+ 0.053\ T - 0.140\ DUM\ 66 + 2.934$$
$$(6.2)(2.6)(16.3)$$

$$\bar{R}^2 = 0.94,\ SE = 0.048,\ DW = 1.7,\ 1956\text{-}1971 \tag{3-41}$$

centrally planned economies

$$\ln PRO = 0.052 \ln PDF_{-3} + 0.231 \ln PDF_{-4} + 0.384 \ln PDF_{-5} + 0.358 \ln PDF_{-6}$$
$$(0.4)(2.0)(3.9)(3.4)$$

$$+ 0.064\ T - 0.281\ DUM\ 63 + 0.177\ DUM\ 66 + 3.496$$
$$(5.8)(3.3)(2.8)(15.3)$$

$$\bar{R}^2 = 0.93,\ SE = 0.060,\ DW = 2.7,\ 1955\text{-}1971 \tag{3-42}$$

For all three groups the estimated relations are consistent with most of the sample period variance in the dependent variable, with no evidence of significant problems from serial correlation.

The long run elasticities of supply implied by the three relations are 0.38 (with a *t*-value of 1.4) for the developed countries, 0.57 (with a *t*-value of 2.4) for the developing countries, and 1.03 (with a *t*-value of 3.2) for the centrally planned economies. The mean lags in the adjustment to changed price expectations for the three groups, respectively, are 0, 4.5, and 5.0 years.

Several features of these price response estimates merit emphasis.

1. These elasticities are relatively large (although for the developed economies the point estimate is significantly nonzero only at the 10

percent level). With the exceptions of cotton and rice for the developed countries, for each country group the price supply elasticity is larger for wheat than for any of the other commodities under study. This result reflects that factors can be shifted in and out of wheat production relatively easily.

2. The ordering of the size of the elasticities across groups is the same as for the other commodities in this study, with the exceptions once again of cotton and rice production in the developed economies. Underlying this ordering in part is the use of relatively specialized inputs in the developed countries which results in less overall flexibility therein. Another important factor apparently is a considerable sensitivity in the centrally planned economies to long run opportunity costs as represented by relative prices. This observation is not meant to suggest that there necessarily are large price responses on a microlevel within such economies, but that the central planners make internal decisions with an awareness of long run opportunity costs in the international markets. They may use price, direct allocation, or other tools to implement those decisions.

3. The rapidity of adjustment to changes in prices across the groups—perhaps because of time needed to change expectations—is in the reverse order of the size of the elasticities. This ordering is roughly consistent with the estimates presented above for other commodities, although for tea and wool the developing economies have a longer lag than do the centrally planned ones. The explanation for this pattern probably lies in the degree of integration in the world market. The developed economies are most integrated, with relatively limited international trade barriers. The developing countries often have substantial trade barriers and attempt to isolate the domestic price of staples such as wheat from the international market. Eventually, however, the pressures in the disequilibrium system build up and adjustments are made. The centrally planned economies also attempt to separate the domestic from the international market. Only when they are sure that international price changes have maintained themselves long enough to represent a longer run opportunity cost do the planners adjust domestic resource allocations.

The nonprice supply determinants include secular trends and dummy variables for special weather conditions. In all three cases the secular trends are quite significant, with annual rates of growth of 3.9, 5.3, and 6.4 percent, respectively. Note that the underlying shifts of resources into wheat and technological changes in wheat production outstripped population growth within each of the three groups. The ordering of these secular trends may reflect the relatively late application of some of the biochemical technological changes in the latter two groups.

The dummy variables, finally, reflect the bad weather conditions in the developed economies in 1961 and 1970, in the developing economies in 1966, and in the centrally planned economies in 1963, as well as the exceptionally good conditions in 1966 for the last of these groups.

Demand

Wheat consumption occurs in considerable degrees in each of the three nation groups. The most important producing group, the centrally planned economies, accounted for 46 percent of total world wheat consumption both in 1956 and in 1971. This level implied imports on the order of magnitude of 1 or 2 percent of world production. Second in importance in wheat consumption was the developed group, which absorbed 32 and 27 percent respectively, in the same two years. Such levels allowed net wheat exports on the order of magnitude of from 3 to 9 percent of world production. Third in importance during the sample period were the developing countries with 22 and 27 percent of world wheat absorption in the same two years. This group expanded its share in absorption at the expense of the developed economies. Since it also expanded its consumption much more rapidly than its output, dependence on net wheat imports increased substantially.

The demand functions for wheat for the three groups of countries are as follows:

developed countries

$$\ln (D/POP) = -0.515 \ln PDF + 0.416 \ln (GDP/POP) - 0.031 \ T$$
$$\quad\quad\quad (4.1) \quad\quad\quad\quad (1.9) \quad\quad\quad\quad\quad (3.6)$$

$$+ 0.288 \ln (PRO/POP) - 2.976$$
$$\quad (3.0) \quad\quad\quad\quad\quad (3.1)$$

$$\bar{R}^2 = 0.71, SE = 0.022, DW = 2.5, 1955\text{--}1971 \quad\quad\quad\quad (3\text{-}43)$$

developing countries

$$\ln (D/POP) = -0.109 \ln PDF_{-0} - 0.157 \ln PDF_{-1} - 0.151 \ln PDF_{-2}$$
$$(1.2) \qquad\qquad (2.1) \qquad\qquad\quad (3.3)$$

$$- 0.096 \ln PDF_{-3} + 0.028 \ \ln (GDP/POP) - 3.234$$
$$(1.6) \qquad\qquad (0.1) \qquad\qquad\qquad (3.1)$$

$$\bar{R}^2 = 0.94, SE = 0.025, DW = 2.8, 1955\text{-}1971 \qquad\qquad (3\text{-}44)$$

centrally planned economies

$$\ln (D/POP) = 0.155 \ln (GDP/POP) + 0.630 \ln (PRO/POP) - 1.490$$
$$(4.1) \qquad\qquad\qquad (7.8) \qquad\qquad\qquad (4.7)$$

$$\bar{R}^2 = 0.93, SE = 0.036, DW = 1.2, 1955\text{-}1971 \qquad\qquad (3\text{-}45)$$

The estimates for the developing and centrally planned economies are quite consistent with variations in the dependent variable over the sample period, although that for the developed economies is less so. In all three cases the Durbin-Watson statistic is in the inconclusive range.

For the developed and the developing countries the price response is relatively large. The respective long run elasticities are −0.51 in both cases (with absolute *t*-values, respectively, of 4.1 and 3.0). These are larger than the point estimates for the same groups for any of the other commodities which are included in this study with the single exception of rice in the developed economies. The mean lags for adjustment of 0 and 1.5 years, respectively, also are at least as rapid as for the other commodities. For these two groups, thus, the demand response to prices is both relatively quick and relatively large.

In the wheat demand relation for the centrally planned economies, on the other hand, there is no evidence of a significantly nonzero price response. This result is in sharp contrast to the other two groups for wheat and to the centrally planned economies price response for other commodities such as cocoa, coffee, and tea. For this basic staple apparently the authorities do not alter the quantities immediately available for consumption because of changes in international prices. They do change internal wheat absorption in a longer run sense, however, by adjusting domestic production which tends to affect internal consumption.

The wheat demand responses to per capita income or product also differ substantially among the three groups and in contrast to those for the other commodities included in this study. The elasticity of 0.4 for the developed countries is second only to that for rice among the income elasticities of this group for nonfiber commodities included in this study. The wheat income elasticity for the developing countries is not significantly nonzero, in contrast to the case for the six commodities above. The wheat income elasticity for the centrally planned economies of 0.15 is significantly different from zero, but also is substantially less than those estimated for cocoa, coffee, and wool (although not for the other products).

Two additional terms are included in the wheat demand relations. For the developed countries there is a secular downward trend of −3.1 percent per year, possibly reflecting a change in diets away from wheat products. For both the developed and the centrally planned economies there are significantly nonzero responses to the level of production internal to the group. The implied elasticities, respectively, are 0.29 and 0.63. A rationale for the inclusion of this variable is provided in the discussion of the demand relations for cotton. In order to maintain relative independence from the international markets, some countries may regulate domestic consumption in relation to production. Only part of any shortfall in domestic production is covered by imports. This rationale well may be valid for the centrally planned economies.[31] It seems much more suspect, however, for the developed economies. Careful consideration of the possibility that this variable might be representing other factors, however, led to no satisfactory alternative explanations.

Price Determination

The estimated relationship for the deflated wheat price[32] is of the form discussed above with one special feature:

$$\ln PDF = -0.164 \ln (STKW/DW)_{-1} - 0.040\ T + 0.037\ \ln IMPC + 0.292$$
$$(3.9) \qquad\qquad (17.4) \qquad (3.8) \qquad\qquad (4.4)$$
$$\bar{R}^2 = 0.97, SE = 0.031, DW = 2.3, 1955\text{--}1971 \qquad\qquad (3\text{--}46)$$

There is a downward secular trend of −4.0 percent per year in the deflated price or, equivalently, of −2.4 percent per year in the proportionality factor between inventories and demand for a given deflated price and given level of wheat imports by the centrally planned economies. As for cocoa, coffee, and wool, this trend probably reflects economies of scale and improved communications and transportation.

The fluctuations in the deflated price around the secular trend are captured fairly well by the lagged ratio of world wheat inventories to world wheat demand. The point estimate of –0.16 implies that the demand for wheat inventories is quite responsive to the deflated wheat price with an elasticity of –6.1 (= –1.0/0.164). This is a greater responsiveness than for the inventory demands for any of the other commodities in this study.[33] This implies that in the wheat market, more than in other markets, random shocks in supply or demand are absorbed much more by inventory adjustments than by price changes.

The final factor in the deflated wheat price determination relation is the import of wheat by the centrally planned economies. In recent years such imports seem to have had an impact beyond that associated with other absorption of wheat, in part because of the visibility and size of some of these deals and in part because of the impact on expectations concerning future flows. The price determination relationship includes a significantly nonzero response to these imports, with an elasticity of 0.04.

The inclusion of this last variable in the price determination relation means that wheat imports of the centrally planned economies enter into the model of the world wheat market. To close the system the following relation is used[34]:

centrally planned economies

$$ln\ (IMP/POP) = -11.347\ ln\ (PRO/POP) + 9.208\ ln\ (D/POP) + 0.142\ T$$
$$(5.1) \qquad\qquad (2.9) \qquad\qquad (3.7)$$

$$-13.071$$
$$(3.4)$$

$$\bar{R}^2 = 0.87, SE = 0.419, DW = 1.9, 1955\text{-}1971 \qquad\qquad (3\text{-}47)$$

This relation is fairly consistent with variations in the dependent variable over the sample. It implies a secular increase of such imports, on a per capita level, of 14.1 percent per year. Imposed on this substantial upward trend are quite elastic responses to internal wheat production and wheat consumption. Price responses enter only indirectly through their lagged effects on internal wheat production.

RICE[35]

Rice is a basic staple in much of the developing and centrally planned world, and production is concentrated in these two groups. The developing countries increased their share in world production from 54 percent in 1956 to 56 percent in 1971. This relatively rapid growth in substantial part reflects the adoption of new high yield varieties, some of which were associated with the "Green Revolution" of the late 1960s. Over the same period the other two groups of economies decreased somewhat their relative shares—from 8 to 6 percent for the developed and from 38 to 37 percent for the centrally planned economies.

Supply

The estimated supply functions for each of the three groups are:

developed economies

$$\ln PRO = 0.206 \ln PDF + 0.152 (\ln PDF_{-0} - \ln PDF_{-1}) + 0.867 \ln PRO_{-1}$$
$$\quad (2.1) \qquad\qquad (1.2) \qquad\qquad\qquad\qquad\qquad (7.8)$$

$$+ 0.361$$
$$\quad (1.2)$$

$$\bar{R}^2 = 0.85, SE = 0.039, DW = 2.0, 1957\text{-}1971 \qquad\qquad (3\text{-}48)$$

developing economies

$$\ln PRO = 0.019\ T - 0.101\ DUM\ 576566 + 0.323 \ln PRO_{-1} + 2.756$$
$$\quad (5.5) \qquad (7.7) \qquad\qquad\qquad (2.8) \qquad\qquad (6.1)$$

$$\bar{R}^2 = 0.98, SE = 0.020, DW = 2.1, 1955\text{-}1971 \qquad\qquad (3\text{-}49)$$

centrally planned economies

$$\ln PRO = 0.118 \ln PDF_{t-2} + 0.110 \ln PDF_{-3} + 0.042 \ln PDF_{-4}$$
$$\quad (1.4) \qquad\qquad (2.4) \qquad\qquad (0.9)$$

$$-0.017 \ln PDF_{-5} + 0.026\ T + 3.623$$
$$\quad (0.2) \qquad\qquad (6.6) \qquad (45.5)$$

$$\bar{R}^2 = 0.96, SE = 0.024, DW = 2.0, 1957\text{-}1971 \qquad\qquad (3\text{-}50)$$

These estimates are quite consistent with variations in the dependent variable for the last two groups, although somewhat less so for the developed countries.

The extent of estimated rice supply response to the deflated price varies considerably across the three groups. For the developed countries the short run price elasticity is 0.36 and long run one is 1.54. These are fairly considerable in comparison to most of the supply price elasticities discussed above for this group of countries. Moreover there is no evidence of a significantly nonzero response in the developed economies to other variables. The adjustment to the long run response, however, is relatively slow, with a mean lag of 6.5 years because much of the production is in special irrigated areas which cannot be adapted easily to other crops.

In the developing countries, in contrast, there is no evidence of a significantly nonzero price response.[36] Instead there is only a distributed lag response to a long run secular change at a rate of 2.8 percent per year and to the poor Asian monsoons of 1957, 1965, and 1966. The secular growth, which reflects both expansion of inputs and technological change, did not exceed population growth by much. Due to income increments, rice demand grew even faster and net rice imports to this group of countries increased substantially.

Rice supply in the centrally planned economies combines both a price response and a secular trend. There is no evidence of a short run significantly nonzero price response. In the long run, however, the estimated price elasticity is 0.25, with a *t*-value of 3.6 and a mean lag of 2.6 years. Although this mean lag is less than for the developed countries, the elasticity for the latter is greater after any given time lapse. The price response in the centrally planned economies is imposed on a secular growth rate of 2.6 percent per year. This secular trend is not significantly different from that for rice production in the developing economies, but has outpaced the slower population growth by somewhat more.

Demand
Although the developed economies became more important net rice exporters to the developing economies over the sample period, this was a change concentrated at the margin. The basic distribution of rice demand among the three groups remained more or less the same as the distribution of production. The developed countries' share dropped from 8 percent in 1956 to 6 percent in 1971, that for the developing countries increased from 54 to 56 percent, and that for the centrally planned economies declined from 38 to 37 percent.

The demand relations for the three groups of countries are:

developed economies

$$ln\,(D/POP) = -0.088\,ln\,PDF_{-0} - 0.169\,ln\,PDF_{-1} - 0.221\,ln\,PDF_{-2}$$
$$\quad\quad (1.5) \quad\quad\quad\quad (2.5) \quad\quad\quad\quad (4.4)$$

$$- 0.224\,ln\,PDF_{-3} - 0.157\,ln\,PDF_{-4} + 2.117\,ln\,(GDP/POP)$$
$$(5.4) \quad\quad\quad\quad (4.3) \quad\quad\quad\quad (2.5)$$

$$- 0.106\,T - 11.851$$
$$(3.3) \quad\quad (3.6)$$

$$\bar{R}^2 = 0.894, SE = 0.036, DW = 2.2, 1957\text{-}1971 \quad\quad\quad (3\text{-}51)$$

developing economies

$$ln\,(D/POP) = -0.007\,ln\,PDF_{-0} - 0.008\,ln\,PDF_{-1} - 0.006\,ln\,PDF_{-2}$$
$$(1.9) \quad\quad\quad\quad (2.5) \quad\quad\quad\quad (2.2)$$

$$- 0.002\,ln\,PDF_{-3} + 0.038\,ln\,(GDP/POP) - 0.012\,DUM\,6566$$
$$(0.8) \quad\quad\quad\quad (5.1) \quad\quad\quad\quad (4.4)$$

$$+ 0.920\,ln\,(PRO/POP) - 0.389$$
$$(48.3) \quad\quad\quad\quad (5.9)$$

$$\bar{R}^2 = 0.998, SE = 0.002, DW = 2.3, 1956\text{-}1971 \quad\quad\quad (3\text{-}52)$$

centrally planned economies

$$ln\,(D/POP) = 0.978\,ln\,(PRO/POP) - 0.00981\,DUM\,59 - 0.0694$$
$$(32.5) \quad\quad\quad\quad\quad (1.9) \quad\quad\quad\quad (0.8)$$

$$\bar{R}^2 = 0.99, SE = 0.0443, DW = 2.0, 1957\text{-}1971 \quad\quad\quad (3\text{-}53)$$

These relations are quite consistent with variations in the dependent variables, although once again somewhat less so for the developed economies than for the other two groups. In this case this aspect of the results probably reflects the role of available supplies within the region in determining absorption for the last two groups. In none of the three estimates does serial correlation appear to be a problem.

For the developed and developing economies, a distributed lag over the deflated rice price has significantly nonzero coefficients. For the former the long run price elasticity is –0.86, with a *t*-value of 4.4 and a mean lag of 2.2 years. The size of this elasticity suggests that considerable substitution occurs between rice and other foods in the long run, although adjustment is not very quick.

For the developing economies, the long run price elasticity is –0.02, with a *t*-value of 2.7 and a mean lag of 1.2 years. This very low price elasticity reflects the dominance of rice and the lack of perceived alternatives in many parts of the developing world, together with the fact that much of the rice is grown and consumed on the same farms.[37] An added factor is that internal rice prices are often fairly isolated from the international market by domestic and foreign sector policies designed to keep wage goods cheap in certain areas.

For the centrally planned economies, however, there is no evidence of a significantly nonzero rice demand response to prices. This result apparently reflects the prevalence to an even greater degree of some of the conditions just described for the developing economies.

For both the developed and developing economies, but not for the centrally planned ones once again, there is evidence of a significantly nonzero response to per capita income or *GDP*. For the developed countries the estimated elasticity is 2.1, which again suggests that rice is hardly a basic staple in most of the relevant areas. The rapidly expanding demand for rice which would occur because of this high elasticity, *ceteris paribus*, is offset by the negative secular trend of –11.0 percent per year—perhaps reflecting a shift in tastes.

For the developing countries the estimated elasticity is 0.04. Despite the consumption of rice almost to the satiation point for some income classes in some regions, this estimate prima facie appears low given the switch from cheaper grains to rice among poor people when income rises in such areas as the Indian subcontinent. In the process of aggregation, however, apparently this latter effect is largely canceled out by the shift away from rice as income rises for higher income classes.

In the developing and centrally planned economies near autarky in production of rice and other basic staples is a widespread political goal. In both cases perhaps the most important determinant of rice absorption, therefore, is the availability of locally produced rice. The implied elasticities of rice consumption with respect to local production are, respectively, 0.92 and 0.96. In both cases severe shortfalls in domestic production, such as due to the failure of the Asian monsoon in the developing world in 1965 and 1966, result in even sharper restraints on consumption because of the depletion of stocks.

Price Determination
For rice the most satisfactory price[38] estimates are obtained with a

geometric adjustment to world stock and demand levels, with modifications for the special conditions in 1967–1968 and 1970:

$$ln\ PDF = -0.337\ ln\ STKW + 0.246\ ln\ DW + 0.118\ DUM\ 6768 - 0.270\ DUM\ 70$$
$$\qquad\quad (1.8) \qquad\qquad\quad (0.9) \qquad\qquad (3.3) \qquad\qquad\quad (4.8)$$

$$+\ 0.919\ ln\ PDF_{-1} - 0.221$$
$$\quad (7.0) \qquad\qquad (0.2)$$

$$\bar{R}^2 = 0.95, SE = 0.0397, DW = 1.7, 1957\text{-}1971 \qquad\qquad (3\text{-}54)$$

This relation is reasonably consistent with variations in the dependent
variable over the sample period. Because of the inclusion of the
lagged dependent variable, of course, the Durbin-Watson statistic is
biased toward two.

The sensitivity of the deflated rice price to rice stock and demand
levels is quite small in the short run with current period elasticities,
respectively, of –0.34 and 0.25. In the long run, on the other hand,
the same elasticities are –4.16 and 3.04. Adjustment to the long run
is not very quick, with a mean lag of 11.3 years! The implications of
these estimates include that in the short run adjustment is primarily
in inventories and not in the price level, but that in the long run the
reverse is true. Given the relatively small price elasticities in the
dominant producing and consuming areas—i.e., excluding the developed countries—the implied feedback through prices in the rice
market is generally both very limited and very slow.

This basic relation between relative rice inventories and the deflated rice price is modified by some special considerations near the
end of the sample period. The failure of the Asian monsoon in 1965
and 1966 caused a considerable depletion in rice stocks. The levels
reached were so low relative to demand that in the two years thereafter there was an added impact on price beyond the log-linear one.
In other words, the demand for rice inventories apparently is more
nonlinear than captured by the log-linear formulation for very low
inventory levels. By 1969, however, expectations had been changed
to be much more optimistic due to the Green Revolution. In that
year, therefore, the previously described effect no longer was relevant and in the next year the changed expectations caused by the
Green Revolution resulted in a significant drop in rice prices below
the level otherwise predicted by relation (3–54). Subsequently the
initial euphoria of the Green Revolution seemed to wear off, so
that further special adjustments in this relation are not required.

The major problem with this price determination relation for rice is the suspiciously slow long run adjustment. As in the case of tea discussed above, therefore, a large number of alternative specifications were explored. The estimated very slow adjustment, however, was quite robust over the alternative specifications which were explored.

NOTES

1. For cocoa, however, uses such as the manufacture of chocolate are more important than its use as a beverage.

2. Ordinary least squares estimation techniques are utilized, with geometric and Almon lags. In the presentation of the results the absolute values of t-statistics are given in parentheses beneath the point estimates, $\bar{R}^2$ is the corrected coefficient of determination, SE is the standard error of the estimated relation, and DW is the Durbin-Watson statistic (which is biased toward two in cases in which the lagged dependent variable is included). The sample period on which the estimates are based is also shown under each equation. For log-linear relationships the coefficient estimates, of course, are elasticities. The variable names used are the same as those given in the previous chapter except dummy variables are indicated by "DUM" with the years for which they have nonzero values subsequently indicated. "ln" is an operator for the natural logarithm.

3. Earlier studies of cocoa, on which in part this report draws, include Acquah (1972), Amoa (1965), Bateman (1965a, 1965b), Behrman (1965 and 1968a), Goreux (1972), Kofi (1972), Mathis (1969), and Weymar (1968).

4. Cocoa trees thrive only within 20 degrees of the equator and most of the world production is within 10 degrees. A mean shade temperature of approximately 80 degrees Fahrenheit with variations not more than ±15 degrees, a well-distributed rainfall of at least 50 inches annually, an altitude between a few hundred feet and a thousand feet above sea level, and protection from strong winds are all usually required. A firm estimate of the percentage of the suitable area which is already under cocoa cultivation, however, is not available. See United Nations—FAO (1965).

5. Behrman (1968a) presents evidence of a significantly nonzero inverse long run response in cocoa supply to the price of competitive crops (i.e. coffee) in half of the country level supply regressions for the eight leading producers during 1948–1964. The exclusion of prices for relevant competitors would bias the estimates of the coefficient of the cocoa price toward zero if the prices of competitors are positively correlated with those of cocoa (and vice versa if the correlations are negative).

6. For the earlier studies (e.g. Bateman [1965a and 1965b] and Behrman [1968a]), internal cocoa prices were used at least for the largest producing countries.

7. Some of the technological change in cocoa production occurred relatively quickly at least in individual countries. For example, spraying spread rapidly in Ghana in the late 1950s and fertilizer use spread rapidly in Brazil in the early 1970s. However, on a worldwide basis the assumption that these changes can be captured largely by the time variable is not inappropriate.

8. We may be overestimating consumption in the developing countries since part of their "grindings," which are considered consumption here, may be exported.

9. Once again, however, part of this growth in the developing countries is growth in grindings for export, not in final consumption.

10. As discussed in the previous section, the price of cocoa here is deflated by the OECD GDP deflator. In addition to the deflated price of cocoa, Behrman (1968a) includes the deflated price of complements and substitutes (i.e. sugar and vegetable oil). For all but one of the eight leading consuming countries during 1948–1964, significantly nonzero coefficients for at least one of these prices are obtained. The exclusion of the price of relevant substitutes (complements) should result in a bias toward (away from) zero in the coefficient estimate for the price of cocoa (and vice versa if it is negatively correlated).

11. The NBER (National Bureau of Economic Research) is publishing a series of books which examine the impact of quantitative restrictions on developing countries. The series is entitled *Foreign Trade Regimes and Economic Development* and is being published by Columbia University Press in 1975 and 1976. The individual volumes are *Turkey* by Anne O. Krueger; *Ghana* by J. Clark Leith; *Israel* by Michael Michaely; *Egypt* by Bent Hansen and Karim Nashashibi; *The Philippines* by Robert E. Baldwin; *India* by Jagdish N. Bhagwati and T.N. Srinivasan; *South Korea* by Charles R. Frank, Jr., Kwang Suk Kim, and Larry Westphal; *Chile* by Jere R. Behrman; *Colombia* by Carlos F. Diaz-Alejandro; *Brazil* by Albert Fishlow; *Anatomy and Consequences of Exchange Control Regimes* by Jagdish N. Bhagwati; and *Liberalization Attempts and Consequences* by Anne O. Krueger.

12. Behrman (1968a) presents similar results for the United States for an earlier time period. For the other leading cocoa consumers among developed mixed economies, however, that study presents evidence of some significantly nonzero income responses.

13. The price series making up the UN export price index for cocoa are:

Country	Specification	Basis of Quotation	Weight Percent
Brazil	Bahia cocoa	Spot, New York	14.5
Cameroon	Fair fermented	c.i.f. Le Havre, nearest forward shipment	12.2
Ghana	Cocoa beans	Spot, New York	24.4
Ghana	Main crop, graded cocoa	c.i.f. U.K., nearest forward shipment	18.9

continued

Country	Specification	Basis of Quotation	Weight (Percent)
Nigeria	Main crop, graded cocoa	Weighted average of f.o.b. prices	18.9
Ivory Coast	Fair fermented	f.o.b. Le Havre, including taxes	11.1

Source: United Nations (1970).

14. Previous econometric studies of coffee include Arak [1969], Bacha [1968], Epps [1970], Wickens, Greenfield, and Marshall [1971], Wickens and Greenfield [1973] and Wilson [1973].

15. Of course, if such a cycle exists on a local or regional basis, it might be swamped out by the aggregation of our data.

16. Also as in the case of cocoa, the relative importance of processing in the centrally planned and developing regions as opposed to the developed region increased during the sample period.

17. Reference here is to the adjustment to the relative price terms without reference to the logarithmic price differences. If the latter are included the ordering is the same, but slower adjustments are implied for both the developed and the developing countries.

18. The Price Series included in the UN export price index for coffee are:

Country	Specification	Basis of Quotation	Weight (Percent)
Angola	Ambriz, no. 2 AA	Spot, New York	5.3
Brazil	Parana type 4	Spot, New York	11.2
Brazil	Santos type 4	Spot, New York	28.4
Brazil	Santos type 2/3	Spot, New York	11.2
Mexico	High grown	Spot, New York	2.7
Venezuela	Tachira, fine washed	Spot, New York	1.3
Colombia	MAMS	Spot, New York	16.0
Ethiopia	Djimmas UGQ	Spot, New York	2.7
Guatemala	Prime washed	Spot, New York	4.0
Uganda	Native Robusta, unwashed, standard grade	f.o.b. Mombasa	9.2
Ivory Coast	Robusta, grade 2, superior	Ex. warehouse, Le Havre	8.0

Source: United Nations (1970).

19. Murti (1961) presents an earlier econometric study of tea.

20. The price indexes included in the UN price index for tea are:

Country	Specification	Basis of Quotation	Weight (Percent)
Sri Lanka	Leaf, high grown	Colombo auctions including export duty and excise	34.4
Sri Lanka	Leaf, all types	Average price, London auctions, less import duty	10.8

Country	Specification	Basis of Quotation	Weight (Percent)
India	Leaf, all types	Average price, Calcutta auctions including export duty and excise	27.9
India	Leaf, North India	London auctions, less import duty	16.9
India	Leaf, South India	Average price, London auctions, less import duty	10.0

Source: United Nations (1970).

21. See for example, Duane (1971), Durbin (1969), McKenzie (1966), and the elaborate econometric model by Witherell (1967).

22. It is necessary to recognize the reversal of the seasons in the lags in these equations since the bulk of the production is in the southern hemisphere.

23. The price series included in the UN export price index for wool are:

Country	Specification	Basis of Quotation	Weight (Percent)
Australia	Wool	Export price index, f.o.b.	51.8
New Zealand	Wool	Export price index, f.o.b.	25.9
South Africa	Type 12, greasy	Average auction price, Port Elizabeth	8.2
Argentina	Crossbreeds 5s/6s (40s/36s) clean basis	In bond, Boston	10.6
Uruguay	Montevideo Super 0s (58s/60s) clean basis	In bond, Boston	3.5

Source: United Nations (1970)

24. See Kolbe and Timm (1971) and Shayal (1960).

25. The series included in the UN price index for cotton are:

Country	Specification	Basis of Quotation	Weight (Percent)
UAR	Menouf, fully good	c.i.f. Liverpool	17.5
Mexico	Tampico-Altamira, M 17/16"	Ex-warehouse, Brownsville Texas	31.3
Turkey	Izmir Standard II	Export price	13.7
USA	Middling 15/16"	Fourteen market average	37.5

Source: United Nations (1970).

26. Law (1975) concludes that commodity agreements and other market interventions were ineffective for price stabilization.

27. Similarly Tewes (1975) found longer lags for cane sugar acreage than for beet sugar.

28. Tewes (1972), presents a nonlinear equation for changes in sugar consumption. While there is a highly significant price effect, we cannot compute elasticities in order to make a comparison. There is no income or population

variable in his equation, but the constant of the equation, which is formulated in terms of changes, accounts for a linear upward trend. The equation fits well in some years, but apparently very badly in others.

29. The component sugar prices and their weights are:

Country	Specification	Basis of Quotation	Weight (Percent)
UK/Cuba	Raw, 96°	Cuban, c.i.f. UK, nearest forward delivery	35.2
UK/Commonwealth	Raw, 96°	Commonwealth Sugar Agreement negotiated price f.o.b. basis, stowed	11.8
Taiwan	Sugar	f.o.b. export price	4.0
United States	Raw 96°	Contract no. 8, f.o.b. and stowed greater Caribbean	18.0
United States	Raw, 96°	Contract no. 10, less duty and freight from Caribbean	31.0

Source: United Nations, (1970)

30. Other econometric studies of wheat include Holt (1955), Mo (1968), and Vannerson (1969).

31. Even for the centrally planned economies the level of aggregation used raises some questions because the rationale might be valid on the country level, but not on the level of all of the centrally planned economies considered together.

32. The component prices of the UN export price index for wheat are:

Country	Specification	Basis of Quotation	Weight (Percent)
Argentina	Up River	c.i.f. U.K., less freight	1.0
Argentina	No. 2 Semihard wheat	f.o.b. Buenos Aires	4.2
Australia	Wheat	Export price index, f.o.b.	14.4
Canada	No. 2 Northern Manitoba	In store, Fort William, Port Arthur	15.3
Canada	No. 3 Northern Manitoba	In store, Vancouver	6.4
Canada	No. 3 C.W. Amber Durum	In store, Fort William, Port Arthur	1.9
Canada	No. 4 Northern Manitoba	In store, Vancouver	2.5
Canada	No. 5 Wheat	In store, Vancouver	1.7
USA	No. 1 Hard winter	f.o.b. Galveston	29.5
USA	No. 1 Soft white	f.o.b. Portland	11.4
USA	No. 2 Soft red winter	f.o.b. Baltimore	4.5
France	Fair, average quality	c.i.f. U.K.	7.2

Source: United Nations (1970)

33. The case of sugar, when inventory levels are far from their minimum levels, is an exception to this statement given the nonlinear nature of relation 3-39.

34. If stock data were available for the centrally planned group, imports could be calculated by subtracting internal production from internal consumption plus internal inventory change.

35. Previous econometric studies of rice include Behrman (1968b), Nasol (1971), and Holder, Shaw and Snyder (1970), Tsujii (1973).

36. Studies such as Behrman [1968b] do present evidence of significantly nonzero price responses in rice production in the developing economies on a much less aggregated level.

37. This latter fact does not necessarily mean that the price elasticity of the marketed surplus of rice is very low. To the contrary, it might be quite high. See Behrman (1968b).

38. The series making up the UN export price index for rice are:

Country	Specification	Basis of Quotation	Weight (Percent)*
Burma	Ngasein, 42 percent broken parboiled	f.o.b. Burma, government contract with Ceylon	27.4
Vietnam	White, no. 1, husked	f.o.b. Saigon	13.1
Italy	Semirough	c.i.f. Germany	9.5
Thailand	White, 5 percent broken	c.i.f. Japan, contract	11.0
Thailand	White, 5 percent broken	f.o.b. Bangkok	8.2
Thailand	White, 5–7 percent broken, fair, average quality	c.i.f. U.K.	8.2
USA	Milled no. 2, medium grain	f.o.b. New Orleans	22.6

*The weights of Cambodia and the U.A.R. have been added to those of Vietnam and Italy respectively.

Source: United Nations (1970).

Chapter 4

Econometric Model Performance over the Sample Period

One test of the performance and realism of econometric models is their ability to describe the operation of the market over the sample period. Even if each of the equations has achieved a good degree of fit, the interrelationships between the equations may or may not provide a realistic prediction of market performance if the model is operated as a simultaneous dynamic system. In this chapter we consider within sample period performance of the models by examining dynamic simulations. The details of these simulations are discussed for each commodity below. The sample period performances generally reinforce our confidence in the use of models presented in Chapter Three for commodity market analysis.

For these simulations, the nonlinear models are solved by a Gauss-Seidel iterative procedure. The simulations are dynamic in that for the nth simulation period lagged endogenous variables from the $n - i$ previous periods are presented not by lagged actual values, but by lagged simulated values.[1] The results of these sample period simulations are summarized for each commodity in Tables 4–1 and 4–2. Table 4–1 shows the average percentage absolute errors of the simulations (error as a percent of actual values) over the sample period. Table 4–2 shows the results year by year.[2] The time paths of the simulations are charted in Appendix C.

Table 4-1. Sample Period Simulation Errors (average percent absolute errors)

	Cocoa	Coffee	Tea	Wool	Cotton	Sugar	Wheat	Rice
World Supply	4.2	4.4	2.2	1.3	3.2	2.5	2.5	1.2
World Demand	2.6	2.6	1.7	2.5	2.6	1.2	2.0	1.0
Price	13.1	4.4	9.2	7.5	4.6	11.3	3.4	3.2
Stocks	11.1	5.9	2.2	5.6	5.7	13.6	14.5	3.0

COCOA

The sample period dynamic simulation for cocoa is satisfactory, although not as good in some dimensions as those for the other commodities presented below. Average percentage errors shown in Table 4-1 are satisfactory for supply and demand, though somewhat high for price and stocks.

The simulations for supply trace the secular developments in cocoa production quite well, but tend to understate the degree of short run variance. The largest percentage errors are overestimates of 12.2 percent in 1958, 7.3 percent in 1973, and 6.8 percent in 1969 and underestimates of –7.1 percent in 1961 and –6.0 percent in 1957. Such errors largely reflect the lack of an adequate specification of the impact of weather changes.

The simulations for world cocoa demands are even better. With the exception of missing the slight decline (i.e., 2.5 percent) in 1958 and indicating that the 1970 trough occurred in 1969, they capture well even relatively minor fluctuations. Among the percentage errors, only the –6.1 percent underestimate for 1957 is as high as 5.0 percent in absolute value.

The simulations for stock, of course, reflect the supply and demand simulations. Although there is a slight one year downturn predicted in 1964 and the 1969 trough is missed by a year, the simulated path generally reflects the major developments in actual stocks quite well. Because the stock levels have averaged much less than annual demand flows during the sample period, however, the above-mentioned supply and demand errors cause larger percentage errors in the stock variable. For example, the stock levels are underestimated by –16.5 percent in 1961 and by –14.4 percent in 1964 and are overestimated by 38.7 percent in 1958, 32.0 percent in 1959, 11.6 percent in 1969, and 36.4 percent in 1973.

The simulated path for the world cocoa price reflects movements in simulated stocks. As with stocks, simulated prices capture the general movements in actual prices fairly well. There is a tendency, however, to predict peaks and troughs one year late (e.g., 1963, 1964, 1969, 1971). Although a large 1973 price increase is simulated, the extent of that increase—over 90 percent—is not fully captured. This latter shortcoming probably reflects the exclusion of price expectations terms. Nevertheless, the general features of the cocoa price simulations are reasonably satisfactory. The relatively large average percentage error is somewhat misleading because the range in the underlying series is relatively large and because the simulation includes the difficult to predict 1973 observation.

COFFEE

The sample period simulation for the coffee model is quite satisfactory. The supply simulations trace the major movements in coffee production quite well, although—as with cocoa—they tend to understate short run variations due to climatic factors. Percentage errors larger in absolute value than 7 percent are encountered only for three of the 18 years—9.9 percent in 1957, –8.0 percent in 1968, and 11.7 percent in 1971.

The mean percentage error for aggregate coffee demand is slightly more.than half of that for supply. Underlying this aggregate error is one of equal percentage size for the developed economies, but one twice as large and one four times as large for the developing and centrally planned economies, respectively. The model thus seems to do best on the demand side with the traditional dominant consuming nations. The model clearly tracks the general movements in aggregate coffee absorption well. Only for three years is the absolute value of the percentage error as large as 5 percent—–5.6 percent in 1963, –5.9 percent in 1966, and 6.8 percent in 1973. Nevertheless, the dynamic simulation does not seem to pick up some slight downturns (e.g., 1958, 1967, and 1973).[3]

The simulated path for stocks reflects the impact of accumulated discrepancies between simulated production and absorption. The historical path is traced out relatively well in the dynamic simulations. Percentage errors as large as 10 percent are found in only two years for coffee—–19.2 percent in 1956 and 18.6 percent in 1957—as contrasted with six years for cocoa. The substantially lower average percentage error for coffee stocks than for cocoa, however, in part

Table 4–2. Summary of Results of Base Simulation (within sample period)*

	1956	1957	1958	1959	1960	1961	1962	1963	1964	1965	—
Cocoa											
Base Simulation											(1973)
Supply	97.01	93.99	112.24	104.70	98.22	92.92	103.10	102.52	95.29	99.65	107.33
Demand	97.77	93.86	101.67	104.93	104.86	101.48	98.80	100.12	101.69	96.89	99.27
Inventories	97.52	97.91	138.70	131.99	102.99	83.50	92.98	98.06	85.56	93.29	136.45
Price	112.11	102.63	80.61	90.40	104.04	114.36	111.49	88.17	108.17	115.00	70.76
Coffee											
Base Simulation											(1973)
Supply	93.03	109.87	101.30	98.68	100.00	98.76	94.57	102.83	94.69	103.63	99.85
Demand	99.22	96.93	102.66	104.57	97.99	102.15	100.00	94.35	99.10	100.52	106.80
Inventories	80.77	118.62	109.51	97.89	100.53	97.41	92.21	99.84	96.01	97.64	93.11
Price	98.75	94.87	95.88	110.50	96.82	102.26	106.69	100.28	101.42	100.36	92.33
Tea											
Base Simulation											(1971)
Supply	102.14	93.16	100.23	101.03	103.62	100.86	102.40	102.35	99.82	100.21	97.89
Demand	100.92	101.26	98.55	98.25	97.94	97.79	99.00	100.30	99.43	101.81	100.95
Inventories	100.38	97.94	98.44	99.24	100.88	101.72	102.66	103.18	103.18	102.66	97.63
Price	103.50	108.84	107.38	106.16	102.05	103.43	95.88	96.71	93.11	90.45	106.10
Wool											
Base Simulation											(1971)
Supply	100.45	103.89	99.04	96.50	97.93	97.17	99.67	99.88	99.49	98.71	102.71
Demand	99.31	102.36	106.03	98.20	95.71	94.09	98.31	97.15	103.81	100.80	103.24
Inventories	103.30	108.07	90.19	86.46	91.69	100.27	104.44	110.13	97.45	90.56	92.19
Price	98.56	94.50	114.54	108.67	105.48	95.44	97.12	96.26	102.28	94.27	125.92

Cotton

Base Simulation										(1970)
Supply	105.02	96.08	101.50	104.38	99.79	96.11	96.62	102.20	101.77	96.75	96.12
Demand	102.51	101.81	100.30	102.21	98.08	94.79	98.53	101.94	104.80	100.80	102.82
Inventories	104.71	94.39	96.34	99.92	103.44	106.84	102.70	103.34	103.20	100.37	78.21
Price	99.23	97.80	96.53	102.33	108.40	96.87	95.11	96.96	99.84	102.84	97.22

Sugar

Base Simulation										(1973)
Supply	102.49	101.93	99.38	97.33	94.92	101.45	97.79	101.76	102.88	104.80	101.75
Demand	103.02	101.45	101.96	99.09	98.45	97.43	96.14	99.39	100.25	100.79	100.38
Inventories	98.08	99.47	86.14	75.49	66.38	84.51	93.17	102.61	118.54	144.77	115.13
Price	80.13	90.01	99.67	132.52	116.39	102.08	93.67	106.57	75.47	93.31	103.80

Wheat

Base Simulation										(1971)
Supply	100.26	106.14	95.07	99.35	99.17	101.80	97.91	101.30	97.58	104.37	96.57
Demand	99.19	102.83	98.15	99.27	101.39	103.59	98.34	98.64	97.35	99.50	95.41
Inventories	105.64	124.70	103.30	103.64	93.52	82.90	81.08	93.66	95.32	134.82	88.70
Price	100.36	100.24	100.00	100.16	101.74	98.45	98.30	103.39	94.08	102.96	108.41

Rice

Base Simulation										(1971)
Supply	100.51	98.70	101.80	98.89	99.64	100.82	99.15	97.69	98.07	99.70	101.65
Demand	100.19	98.44	101.98	99.21	99.49	100.87	99.50	98.00	98.14	99.83	100.79
Inventories	102.12	103.95	102.70	100.36	101.41	101.05	98.37	95.94	95.43	94.53	103.39
Price	97.08	93.92	100.90	104.48	99.78	95.52	97.66	102.42	102.99	102.19	103.98

*simulated values/actual values × 100.0.

reflects that the average stock levels as percentage of annual demands are much higher for coffee (i.e., from 32 to 135 percent during the sample as compared to a maximum of 52 percent for cocoa).

The relative success in simulating coffee stocks and the limited sensitivity of the coffee price to the inventory demand ratio result in quite satisfactory coffee price simulations. Most of the major price movements are well represented although the details of the bottoming out of troughs in the early and mid-1960s are not captured. A large 1973 price increase is simulated, but—once again—not to the degree that it actually occurred. This latter fact probably relates again to the exclusion of price expectations terms in the demand for inventory normalization of the price determination relationship. All in all, nevertheless, the sample period coffee price simulation is quite successful considering the 18 year length of the dynamic simulation and the long lags for feedback on the supply side.

TEA

In many respects the sample period dynamic simulation for tea is quite good. Average percentage errors for the major aggregate variable over the 1956–1971 period shown in Table 4–1 are satisfactory.[4]

The simulations for supply trace the secular developments in tea production quite well, but tend to understate the degree of short run variance that is probably attributable to the weather. The largest percentage error, and the only one as large as 5 percent, is the 6.7 percent underestimate for 1957. The average percentage errors imply greater success for tea than for the previously discussed commodities, although even greater success is encountered in some cases below (e.g., wool and rice).

As is the case for most of the other commodities, the simulations for tea demand are better than those for supply. For either price relation the average percentage errors are low in comparison to the other commodities, with the single exception of rice. The 4.2 percent overestimate for 1968 with the first price relation is the only percentage error as large as 4 percent in absolute value.

Given the relative success of the tea supply and demand simulations and that stocks usually have been equal to the demand flow over several years, it is not surprising that the average percentage errors for inventories also are quite low in comparison to most other commodities included in this study. The largest absolute percentage error—and the only one over 4 percent—is the −4.5 percent shortfall for 1971. The differences between actual and simulated values are highly correlated for both dynamic simulations with negative values for 1956 and 1960 through 1967 or 1968 and positive values otherwise. Such autocorrelation is not surprising since a large random shock in one of the supply or demand flow variables in any one year is carried forward in this stock variable.

The simulated path of the deflated tea price captures quite well the secular decline in the actual price. The average percentage error, thus, is reasonably low.

As is suggested above, however, the model does not seem to pick up the short run fluctuations well. There is considerable serial correlation, moreover, in the difference between actual and simulated tea price values. For 1962 through 1969, to be more explicit, the simulated values systematically underestimate the price level while for all other years they overestimate it. During the mid-1960s, moreover, the percentage errors are quite large—as much as −24.3 percent in 1967 for the first price relation.

The importance of the inventory errors can be investigated by operating the model with inventories as an exogenous variable. For such a simulation the average percentage error in the tea price falls to 4.9 percent. This is better than for the simulation with endogenous stocks. However the same pattern of serial correlation—albeit alternated somewhat—remains. This experiment suggests that the real problem is in the slow price adjustment. Since prices adapt so slowly and price elasticities are so low, moreover, feedback to offset shocks to the inventory level takes a very long time to work itself out. The strong serial correlation in the difference between actual and simulated inventories, therefore, is not the cause of the serial correlation in the difference between actual and simulated tea prices. To the contrary, in fact, it reflects the slow supply and demand responses to surplus (deficit) inventories which occur because of the very slow price adjustment. For these reasons, as is noted above, considerable efforts were made to explore alternative specifications for the price determination relations.

WOOL

The sample period simulation for the wool model is quite satisfactory, with relatively low average percentage errors.

The equations catch the upward movement of supply with considerable accuracy except for a sharp decline in supply from the developed mixed economies in 1971. The errors in 1957 and 1971 may represent inadequacies in the statistics since they are present both in the production and consumption statistics of the developed economies. It is interesting to note that, in contrast to most of the other commodities, supply is modeled more accurately than demand. The model catches the price movements in the wool market with considerable precision.

COTTON

The within sample period simulation for cotton is relatively satisfactory.

The equation system catches the upward movement of supply, although there is considerable unexplained residual variation in some cases. In the developed countries, dominated by United States production and hence by U.S. agricultural policy, the model does not recognize the sharp downward shift of output in 1967 although it does catch the relatively low level of production in subsequent years. The pattern of output is caught well in the developing economies, but there are instances of large deviations of output from the predicted pattern in the centrally planned economies, possibly as a result of weather conditions.

On the demand side the world demand for cotton is captured with considerable precision. In the developed economies the range of variation is quite small from the beginning to the end of the period so that relatively small variations take on an exaggerated importance. It is interesting to observe that the errors noted in the supply in the centrally planned economies have corresponding errors on the demand side. In these economies consumption is often constrained by available supplies, a factor which we note with other commodities as well (e.g. wheat and rice).

The system computation for price is very satisfactory. The model simulates the downward path of price in real terms (*PDF*) with little error. In nominal terms the model records the sharp decline of the nominal price between 1955 and 1959 and catches the upward movement of price in the last year of the simulation period, 1971. This satisfactory record of performance during the sample period may however be deceiving for the postsample period in view of the high autocorrelation of the price equation.

SUGAR

The sample period dynamic simulations are reasonably satisfactory in catching the essential movements of the principal variables—though we should not lose sight of the fact that dummy variables have been introduced in four places in the production equation for the developing countries and in three time periods for the price determination equation. The average percentage simulation errors for the major aggregates over the 1955–1973 sample period are satisfactory for supply and demand but relatively high for price and stocks.

The performance of the supply side of the model is better with regard to explaining the underlying trend than with regard to catching production fluctuations. We have noted the dummy variables in the equation for production in the developing economies. There are also substantial misses in the explanation of supply in other parts of the world. No doubt these problems relate to the dependence of sugar production on weather conditions.

Demand is well explained except for some rather surprising variations in consumption in the centrally planned economies. Despite the problems noted above, the stocks of sugar are quite well explained. This is a consequence of the fact that the unexplained variations in supply are predominantly one year misses reflecting weather conditions, so that there is not substantial cumulative buildup or decline in inventories. The price simulation shows a reasonably good representation of the sugar price. It is, however, unable to capture the wild fluctuations which are part of price movement in the sugar market.

To supplement the deterministic simulations discussed above, which assumed that random errors were at the expected value of zero, we computed a stochastic simulation. This simulation simply added to the supply equations random values drawn from a random normal distribution with the standard deviation corresponding to the standard error of the simulated equations. (For the developing countries the standard error of a supply equation without the dummy weather variables was used.)

A large part of the random variations unexplained in the equation estimation must be attributable to weather variations so that the stochastic simulations show system performance when the weather variations are introduced. Twenty stochastic simulations were run and the simulation errors were computed as the difference between the stochastic simulations and the deterministic simulation above.

The stochastic simulation errors (relative to the deterministic simulation) are very close in magnitude to the errors observed in the deterministic solution (relative to the actual observations). This is not surprising and suggests that much of the error in the equation system originates with unexplained crop variations which are the result of the weather conditions.

WHEAT

The sample period dynamic simulation for wheat is quite satisfactory in most respects. Average percentage errors for the major aggregates over the 1956–1971 sample period are moderate except for stocks.

The simulations for world wheat supply trace the major developments in the wheat market quite well. They capture the troughs in 1961, 1963, 1967, and 1970 and the peaks in 1958, 1962, 1968, and 1971 (although the first and third of these peaks are represented with a year lag). Some slight downturns—such as in 1957 and 1965—are not picked up. Nevertheless, the general success in capturing supply fluctuations, the low average percentage error, and the fact that for only one year does the simulated wheat supply differ from the actual level by as much as 5 percent (i.e., a 6.1 percent overestimate in 1957) all point to relatively great success in this dimension.

The simulations of world wheat demand are even better. They miss the slight downturns in 1960 and 1961 and the almost imperceptible one in 1969, but track all other turning points with the correct timing. The biggest absolute percentage error, the only one as large as 4 percent, is a 4.6 percent underestimate for 1971. The low average percentage error is bettered only by tea, sugar, and rice among the commodities considered.

Given the relatively great success in tracking supply and demand, it is not surprising that the model does comparatively well in predicting changes in stocks. Because inventories are so low relative to demand (from 9.1 to 22.6 percent of the annual demand during the sample period) for wheat, as for cocoa, the average percentage errors are relatively high. Nevertheless, the simulated stocks trace fairly well the general pattern of actual stocks, even though they miss the timing of the 1965 trough and the extent of both the 1965 trough and the 1969 peak.

The simulated deflated wheat price reflects the simulated stock movements. But, as is emphasized above, in the wheat model, in comparison to the others, inventories absorb shocks relatively more than prices. As a result, the simulated price follows the secular trend of the actual price in the wheat model better than in any other model except for rice, with an average percentage error of only 3.4. Still, some difficulties remain in the identification of turning points for the deflated price. Although the simulation captures such major troughs as that in 1971, it lags the 1960 trough and 1962 peak by one year and misses entirely the slight upturns in 1964 and 1966. As a result, in five of the 16 years the percentage error in the deflated price is greater than 5 percent (i.e., shortfalls of –5.9 percent in 1964, –8.4 percent in 1966, and –5.5 percent in 1967 and overestimates of 9.9 percent in 1970 and 8.4 percent in 1971).

RICE

The sample period dynamic simulation for rice is better than for any of the other commodities discussed in this study. Average percentage errors for the major aggregate variables over the 1957–1971 period are the lowest obtained for any commodity, with the single exception of the inventory level for tea.[5]

This simulation traces out both supply and demand quite well. The slight drop in demand in 1959 is missed and a slight increase instead of the actual fall in both supply and demand are predicted in 1966, but in other respects the extent to which the actual paths are captured is striking. For both supply and demand, for example, there is only one year in which the predicted value differs from the actual value by as much as 2 percent—and in that case the discrepancies are less than 2.5 percent.

For rice stocks and the deflated rice price the success is also considerable. For stocks the model misses by more than 5 percent only for the –5.5 and –7.8 percent underpredictions for 1966 and 1967. For prices, likewise, there are only two discrepancies of such a magnitude—a –6.1 percent underprediction in 1958 and a 5.7 percent overestimate in 1968. For deflated rice prices, moreover, the small 1962 peak is missed. Despite these exceptions, however, the general performance of the model in tracing rice stocks and deflated rice prices is quite noteworthy.

NOTES

1. Of course, at the beginning of the simulation actual lagged values are used.

2. The values in each case are: (estimated value/actual value) × 100.

3. It might only be calculating the 1973 drop with a lag of a year, as it does for 1964, but the data were not available to test this possibility.

4. The errors shown are for the version using price equation (3-17). The alternative price relation (3-18) yields slightly lower errors. Version (3-17) has also been used for sample period and multiplier simulations.

5. As is noted above, however, comparisons of the percentage changes in inventories across commodities are, in an important sense, misleading because the ratios of the levels to annual demands vary considerably across commodities.

 *Chapter 5*

Econometric Model
Multiplier Simulations

Commodity markets are complex dynamic simultaneous systems, even in their simplest representation. The behavioral properties of the markets themselves and of models are frequently difficult to predict from analysis of market behavior or from the characteristics of the separate elements which make up supply, demand, and price determination. For this reason dynamic model simulations are an excellent means of testing the performance of the models and, if the models are realistic in their essentials, of the markets themselves.

The responses of econometric model systems to exogenous changes are best measured through multiplier simulations. Since general analytic solutions are difficult or impossible to establish for non-linear dynamic systems such as the commodity market models presented in this study, simulation methods are used to compute the response properties of the models over time to changes in the exogenous variables. These so-called multiplier simulations measure the difference between the base simulation and an alternative simulation which includes a specific exogenous change. In this section, we consider multiplier simulations for the eight commodity market models.

SIMULATION ASSUMPTIONS

Four types of multiplier calculations were carried out and compared to the base solution:

Ia.   A one time reduction in world supply of 5 percent.
Ib.   A one time increase in world demand of 5 percent.
IIa. A continued 1 percent increase in world GDP over levels prevailing in the base solution.
IIb. A continued 1 percent increase in the rate of growth of world GDP.

The pattern of these "disturbances" introduced into the multiplier simulation is shown in Figure 5–1.

The first two multiplier simulations, Ia and Ib, measure the impact of exogenously determined changes in demand and supply on the commodity market. The supply change resembles closely the impact of a year of poor weather, a frequent and important occurrence in these markets, as we have noted above. After the poor harvest, which is assumed to reduce supply by 5 percent, the supply relationships return to their normal level in subsequent years. The demand change may be seen as a one time exogenous increase in demand reflecting a sudden but temporary improvement in economic conditions. Official stockpile purchases can be expected to have somewhat similar effects.

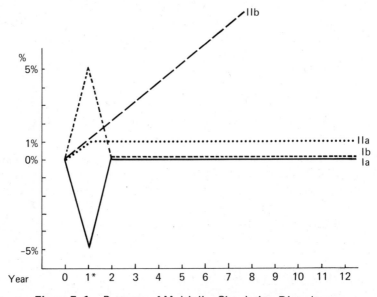

Figure 5–1.   Patterns of Multiplier Simulation Disturbances

*In our calculations year 1 corresponds to 1956 except in the case of cotton where the initial disturbance is in 1955, and rice where it is 1957.

In contrast, the simulations labeled IIa and IIb represent persistent changes in the exogenous variables of the system. Multiplier simulation IIa measures the impact of an improvement in the gross domestic product of the consuming countries amounting to 1 percent. The 1 percent increase has been imposed beginning in the first year and extending throughout the simulation period. Thus each year GDP is just 1 percent higher in the multiplier solution than in the base solution. The question posed by this multiplier calculation is not only to measure the effect of a change in GDP but also to evaluate its impact if the change persists over time. Ultimately, do supply, demand, and inventories adjust sufficiently for price to return to near its base solution level? The second of these simulations poses the same problem only in more severe terms. In multiplier simulation IIb we measure the impact of a persistent change in the rate of growth of GDP, an increase of 1 percent over the entire sample period. This means that by the end of the sample period GDP is 1.01^n times higher than in the base simulation where n is the number of years over which the higher growth rate has been assumed. This simulation thus assumes a considerably higher and persistent change in GDP. It is most interesting to ask how production adjusts to a higher GDP growth rate and whether more rapid growth produces higher prices.

MULTIPLIER SIMULATION RESULTS

The results of the multiplier simulations for the one time changes in demand and supply (simulations Ia and Ib) are summarized in Table 5-1. In each case the figures in Table 5-1 show the multiplier simulation values relative to the base solution values times 100. Thus the value of 101 means that the multiplier solution value is 1 percent higher than the corresponding result in the base solution. The same procedure is used to summarize results of multiplier simulations IIa and IIb in Table 5-2. It is important to remember, however, that in these cases the assumed changes persist over the simulation period and they are only 1 percent in magnitude, on the level of GDP in simulation IIa and on the rate of change of GDP in simulation IIb.

Next we proceed to examine the multiplier simulations for each commodity. The response patterns obtained in the multiplier simulations for the commodities are quite varied. In some instances, the

Table 5-1. Summary of Results of Simulations Ia and Ib*

	1956	1957	1958	1959	1960	1961	1962	1963	1964	1965	—	1973
Cocoa												
Simulation Ia 5 Percent Decrease in Supply												
Effect on:												(1973)
Supply	95.00	100.00	100.00	100.00	100.00	100.00	100.27	101.03	101.84	101.97		98.68
Demand	99.45	98.07	98.63	98.96	99.86	100.08	100.16	100.20	100.51	100.96		99.16
Inventories	87.06	92.89	96.91	100.44	100.74	100.42	100.63	102.33	105.43	106.13		93.95
Price	112.10	104.76	101.52	98.73	99.25	99.71	99.60	98.21	95.98	95.80		104.74
Simulation Ib 5 Percent Increase in Demand												
Effect on:												
Supply	100.00	100.00	100.00	100.00	100.00	100.00	100.34	101.22	102.02	101.94		98.66
Demand	104.27	97.72	98.85	99.09	100.12	100.14	100.16	100.21	100.59	101.07		99.10
Inventories	88.10	94.94	98.38	101.30	100.78	100.31	100.68	102.77	106.14	106.39		93.69
Price	115.57	102.51	100.41	98.12	99.44	99.86	99.56	97.85	95.50	95.69		104.94
Coffee												
Simulation Ia 5 Percent Decrease in Supply												
Effect on:												(1973)
Supply	95.00	100.00	100.00	100.00	100.00	100.00	99.83	100.37	101.11	101.42		99.91
Demand	99.97	97.30	99.88	99.94	99.76	99.96	99.90	99.86	99.96	100.09		100.21
Inventories	82.45	93.77	96.00	97.20	98.57	98.68	98.64	99.24	100.24	101.20		101.84
Price	107.77	101.44	101.55	101.08	100.47	100.50	100.50	100.24	99.89	99.57		99.37
Simulation Ib 5 Percent Increase in Demand												
Effect on:												
Supply	100.00	100.00	100.00	100.00	100.00	100.00	99.80	100.47	100.27	101.50		99.88
Demand	104.96	96.77	100.35	99.98	99.73	99.99	99.91	99.87	99.99	100.12		100.21
Inventories	83.29	95.76	96.48	97.49	98.76	98.83	98.74	99.43	100.54	101.55		101.79
Price	109.40	100.41	101.54	100.99	100.38	100.45	100.46	100.17	99.79	99.45		99.40

Tea

Simulation Ia 5 Percent Decrease in Supply

Effect on:

											(1971)
Supply	95.00	100.12	100.20	100.23	100.29	100.38	100.50	100.62	100.69	100.69	99.89
Demand	99.96	99.89	99.70	99.65	99.66	99.68	99.72	99.78	99.85	99.92	100.23
Inventories	98.56	98.65	98.82	99.01	99.23	99.46	99.70	99.95	100.20	100.41	100.56
Price	101.11	101.86	102.17	102.23	102.10	101.84	101.47	101.02	100.53	100.04	98.68

Simulation Ib 5 Percent Increase in Demand

Effect on:

											(1971)
Supply	100.00	100.51	100.45	100.32	100.49	100.65	100.90	101.01	100.91	100.66	99.41
Demand	104.81	99.67	99.11	99.53	99.69	99.77	99.86	99.96	100.07	100.17	100.27
Inventories	98.77	99.03	99.43	99.66	99.91	100.17	100.47	100.75	100.98	101.09	100.24
Price	104.76	104.26	103.10	102.33	101.67	101.00	100.31	99.63	99.00	98.50	98.74

Wool

Simulation Ia 5 Percent Decrease in Supply

Effect on:

											(1971)
Supply	95.68	98.28	99.81	99.93	99.90	99.83	99.59	99.61	99.80	99.96	99.94
Demand	99.86	95.22	98.83	99.77	99.75	100.14	100.29	99.45	99.46	99.71	99.89
Inventories	88.04	96.92	99.74	100.18	100.61	99.73	97.74	98.28	99.20	99.95	99.65
Price	108.52	93.95	98.20	99.49	99.18	100.42	102.02	100.22	99.61	99.55	100.05

Simulation Ib 5 Percent Increase in Demand

Effect on:

											(1971)
Supply	101.26	98.67	100.30	100.57	100.58	100.45	99.92	99.69	99.77	99.95	99.91
Demand	104.74	96.19	99.06	100.14	100.14	100.79	101.25	100.00	99.57	99.63	99.92
Inventories	90.60	97.83	101.32	102.47	103.73	102.67	98.93	98.15	98.64	99.56	99.34
Price	115.46	94.98	97.55	98.64	97.85	99.59	102.85	101.24	100.18	99.67	100.30

continued

Table 5-1 continued

	1956	1957	1958	1959	1960	1961	1962	1963	1964	1965	—	1973
Cotton												
Simulation Ia 5 Percent Decrease in Supply												
Effect on:												(1970)
Supply	95.00	99.13	99.52	99.83	100.22	100.74	101.45	102.07	102.42	102.59		93.70
Demand	99.03	99.21	99.31	98.13	98.77	98.48	98.47	98.81	99.55	100.67		102.63
Inventories	92.52	92.38	92.76	93.94	96.75	101.29	107.76	115.35	119.26	121.27		49.38
Prices	100.29	101.39	103.64	106.81	109.45	110.93	110.48	107.50	102.06	95.01		85.20
Simulation Ib 5 Percent Increase in Demand												
Effect on:												
Supply	100.00	100.02	100.07	100.25	100.69	101.44	102.46	103.23	103.48	103.37		90.94
Demand	105.00	100.22	99.65	98.99	98.26	97.91	98.13	98.86	100.14	101.82		102.39
Inventories	91.32	90.95	91.71	93.95	98.69	105.82	115.46	125.74	129.31	129.82		17.59
Price	100.60	102.66	106.53	111.41	114.36	115.13	112.74	106.88	98.18	88.35		92.88
Sugar												
Simulation Ia 5 Percent Decrease in Supply												
Effect on:												(1973)
Supply	95.00	100.17	101.95	101.05	100.29	99.89	99.44	99.84	100.31	100.04		100.01
Demand	99.55	99.30	99.56	99.85	100.02	100.03	99.93	99.90	99.97	100.00		100.00
Inventories	83.54	84.56	94.04	101.62	103.06	101.63	99.13	98.89	100.36	100.48		100.06
Price	114.85	106.36	100.29	96.99	97.44	99.42	101.72	100.56	99.15	99.73		99.96
Simulation Ib 5 Percent Increase in Demand												
Effect on:												
Supply	100.00	100.20	102.22	101.52	100.04	99.36	99.52	100.12	100.34	100.06		100.00
Demand	104.48	99.24	99.59	100.02	100.16	100.01	99.89	99.92	99.99	100.02		100.00
Inventories	83.51	84.90	95.84	106.03	104.06	99.68	98.13	99.00	100.50	100.64		100.08
Price	117.14	105.98	98.92	93.40	97.59	101.63	101.85	99.91	99.10	99.65		99.97

Wheat

Simulation Ia 5 Percent Decrease in Supply

Effect on:											(1971)
Supply	95.13	100.34	100.29	100.18	100.29	100.58	100.78	100.62	100.43	100.20	99.90
Demand	97.92	99.63	99.65	99.70	99.87	100.16	100.44	100.51	100.42	100.28	99.85
Inventories	85.78	90.22	93.74	96.03	97.92	99.78	101.98	102.71	102.62	101.94	98.85
Price	101.07	102.19	101.64	100.99	100.54	100.12	99.72	99.41	99.42	99.53	100.31

Simulation Ib 5 Percent Increase in Demand

Effect on:											(1971)
Supply	100.22	100.72	100.49	100.15	100.48	101.06	101.39	101.04	100.49	100.17	99.90
Demand	104.71	99.18	99.48	99.53	99.85	100.41	100.91	100.95	100.60	100.37	99.78
Inventories	77.90	86.86	92.29	95.27	98.13	101.31	104.49	105.13	104.15	102.62	98.34
Price	101.69	104.95	102.19	101.22	100.59	99.90	99.23	98.84	99.06	99.31	100.50

Rice

Simulation Ia 5 Percent Decrease in Supply

Effect on:											(1971)
Supply	95.01	98.81	99.57	100.01	100.36	100.61	100.69	100.58	100.32	99.90	99.47
Demand	95.70	99.15	99.64	99.81	99.92	100.01	100.10	100.16	100.20	100.16	99.71
Inventories	95.69	93.55	93.20	94.56	97.67	102.07	106.67	110.26	111.06	108.97	88.87
Price	100.40	102.44	104.60	106.15	106.45	105.18	102.52	99.04	95.73	93.37	104.36

Simulation Ib 5 Percent Increase in Demand

Effect on:											(1971)
Supply	100.41	100.80	101.62	102.46	102.74	102.17	101.05	99.62	98.11	96.93	102.44
Demand	104.85	99.56	99.67	100.11	100.50	100.59	100.60	100.51	100.33	100.04	100.18
Inventories	71.32	80.22	93.67	109.86	125.65	138.08	142.19	136.04	118.36	93.97	85.97
Price	113.37	120.74	121.45	115.84	106.12	94.87	84.74	77.54	74.85	78.27	142.36

*multiplier simulation values/base simulation values × 100.0.

Table 5-2. Summary of Results of Simulations IIa and IIb*

	1956	1957	1958	1959	1960	1961	1962	1963	1964	1965	— (1973)
Cocoa											
Simulation IIa 1 Percent Increase in GDP											
Effect on:											
Supply	100.00	100.00	100.00	100.00	100.00	100.00	100.01	100.04	100.11	100.20	100.21
Demand	100.11	100.10	100.08	100.04	100.04	100.06	100.07	100.08	100.10	100.13	100.29
In Developed	99.98	99.90	99.83	99.76	99.74	99.74	99.74	99.74	99.73	99.75	99.77
Developing	100.43	100.70	100.88	100.99	101.06	101.11	101.15	101.17	101.19	101.20	101.25
Centrally Planned	101.13	101.05	100.91	100.77	100.69	100.66	100.67	100.68	100.67	100.70	100.73
Inventories	99.69	99.40	99.09	99.08	99.12	99.15	99.15	99.14	99.10	99.56	99.15
Price	100.37	100.61	100.86	100.83	100.79	100.79	100.80	100.82	100.86	100.49	100.99
Simulation IIb 1 Percent Increase in Rate of Growth of GDP											
Effect on:											
Supply	100.00	100.00	100.00	100.00	100.00	100.00	100.01	100.05	100.17	100.40	103.00
Demand	100.11	100.22	100.31	100.36	100.44	100.58	100.72	100.86	100.96	101.12	103.88
In Developed	99.98	99.88	99.70	99.42	99.18	98.93	98.69	98.41	98.00	97.58	94.81
Developing	100.43	101.13	102.01	103.01	104.10	105.26	106.48	107.73	108.99	110.23	121.02
Centrally Planned	101.13	102.19	103.09	103.86	104.55	105.25	105.99	106.68	107.14	107.80	111.05
Inventories	99.69	99.05	97.93	97.13	96.44	95.74	94.91	93.72	91.39	92.84	80.68
Price	100.37	101.01	102.09	102.85	103.54	104.32	105.22	106.50	108.93	107.61	124.22
Coffee											
Simulation IIa 1 Percent Increase in GDP											
Effect on:											
Supply	100.00	100.00	100.00	100.00	100.00	100.00	99.99	100.01	100.05	100.08	100.09
Demand	100.16	100.05	100.09	100.10	100.10	100.12	100.10	100.09	100.10	100.09	100.05
In Developed	100.20	100.08	100.15	100.14	100.14	100.16	100.13	100.13	100.14	100.13	100.08
Developing	100.00	99.96	99.85	99.93	99.92	99.93	99.94	99.91	99.91	99.92	99.87
Centrally Planned	101.28	101.13	101.15	101.14	101.17	101.18	101.13	101.12	101.12	101.10	100.88
Inventories	99.45	99.45	99.46	99.47	99.60	99.51	99.43	99.43	99.42	99.36	98.86
Price	100.28	100.24	100.24	100.25	100.20	100.24	100.26	100.26	100.26	100.29	100.47

Simulation IIb 1 Percent Increase in Rate of Growth of GDP

Effect on:

Supply	100.00	100.00	100.00	100.00	100.00	100.00	99.99	100.00	100.05	100.12	100.69
Demand	100.16	100.21	100.33	100.46	100.60	100.80	100.84	100.95	101.06	101.11	100.85
In Developed	100.20	100.27	100.45	100.61	100.76	100.99	101.05	101.18	101.35	101.46	101.38
Developing	100.00	99.96	99.82	99.79	99.74	99.71	99.70	99.52	99.45	99.39	97.66
Centrally Planned	101.28	102.45	103.72	105.00	106.40	107.72	108.86	110.09	111.34	112.42	116.80
Inventories	99.45	99.05	98.75	98.44	98.55	97.91	97.22	96.81	96.19	95.08	80.33
Price	100.28	100.45	100.62	100.79	100.80	101.13	101.43	101.64	101.94	102.42	109.24

Tea

Simulation IIa 1 Percent Increase in GDP

Effect on:

											(1971)
Supply	100.00	100.01	100.03	100.04	100.05	100.07	100.10	100.13	100.16	100.18	100.17
In Developed	100.00	100.00	100.00	100.00	100.01	100.02	100.03	100.04	100.05	100.05	100.03
Developing	100.00	100.02	100.04	100.05	100.06	100.07	100.08	100.11	100.13	100.15	100.14
Centrally Planned	100.00	100.00	100.00	100.00	100.03	100.09	100.17	100.24	100.30	100.34	100.30
Demand	100.13	100.13	100.11	100.09	100.08	100.08	100.08	100.08	100.08	100.09	100.15
In Developed	99.99	99.98	99.97	99.97	99.97	99.96	99.96	99.96	99.96	99.97	99.99
Developing	100.43	100.43	100.39	100.37	100.36	100.35	100.35	100.35	100.35	100.36	100.40
Centrally Planned	100.00	99.96	99.92	99.87	99.83	99.80	99.78	99.77	99.77	99.77	99.89
Inventories	99.97	99.94	99.92	99.90	99.90	99.91	99.92	99.94	99.97	100.00	100.12
Price	100.13	100.25	100.35	100.43	100.48	100.52	100.53	100.53	100.50	100.46	100.19

Simulation IIb 1 Percent Increase in Rate of Growth of GDP

Effect on:

Supply	100.00	100.01	100.04	100.08	100.13	100.21	100.31	100.45	100.61	100.80	102.08
In Developed	100.00	100.00	100.00	100.00	100.01	100.03	100.07	100.11	100.16	100.21	100.49
Developing	100.00	100.02	100.06	100.11	100.18	100.25	100.34	100.45	100.59	100.75	101.85
Centrally Planned	100.00	100.00	100.00	100.00	100.03	100.13	100.30	100.56	100.87	101.23	103.58
Demand	100.13	100.27	100.39	100.48	100.57	100.66	100.75	100.84	100.93	101.04	101.90
In Developed	99.99	99.97	99.94	99.91	99.88	99.84	99.80	99.76	99.72	99.68	99.50
Developing	100.43	100.86	101.25	101.62	101.98	102.34	102.70	103.06	103.41	103.78	106.10
Centrally Planned	100.00	99.96	99.88	99.74	99.57	99.36	99.13	98.89	98.65	98.41	97.21
Inventories	99.97	99.90	99.81	99.71	99.61	99.51	99.43	99.37	99.33	99.32	99.67
Price	100.13	100.39	100.77	101.22	101.72	102.27	102.85	103.42	103.99	104.54	107.15

continued

Table 5–2 continued

	1956	1957	1958	1959	1960	1961	1962	1963	1964	1965	—	1973
												(1971)
Wool												
Simulation IIa 1 Percent Increase in GDP												
Effect on:												
Supply	100.45	99.99	100.07	100.30	100.52	100.68	100.65	100.53	100.48	100.46		100.57
In Developed	100.73	99.98	99.98	100.05	99.99	100.02	100.19	100.14	100.14	100.09		100.11
Developing	100.00	100.00	100.14	100.72	101.56	102.09	101.49	101.11	100.90	100.85		101.12
Centrally Planned	100.00	100.05	100.26	100.66	101.10	101.24	101.25	101.11	101.13	101.15		101.36
Demand	101.70	100.33	99.95	100.05	100.11	100.35	100.78	100.78	100.65	100.45		100.61
In Developed	101.81	99.91	99.15	99.23	99.33	99.80	100.62	100.58	100.34	99.94		100.10
Developing	100.04	99.68	100.26	101.09	101.03	101.01	100.79	100.49	100.39	100.51		100.55
Centrally Planned	102.06	102.06	102.06	102.07	101.85	101.48	101.21	101.45	101.61	101.71		101.57
Inventories	96.61	95.54	96.23	96.99	98.11	99.11	98.72	98.16	97.46	97.12		97.65
Price	105.28	103.63	102.48	102.12	101.46	101.18	102.19	102.57	102.84	102.72		102.63
Simulation IIB 1 Percent Increase in Rate of Growth of GDP												
Effect on:												
Supply	100.45	100.48	100.47	100.85	101.40	102.01	102.78	103.19	104.07	104.47		108.73
In Developed	100.73	100.75	100.64	100.80	100.81	100.67	101.00	100.91	101.58	101.40		102.07
Developing	100.00	100.00	100.14	100.87	102.47	104.64	106.25	107.50	108.61	109.51		118.58
Centrally Planned	100.00	100.05	100.32	100.99	102.11	103.41	104.74	106.05	107.29	108.53		118.50
Demand	101.70	102.05	101.90	102.05	102.16	102.35	103.19	103.80	104.93	104.57		108.73
In Developed	101.81	101.73	100.60	100.05	99.12	98.45	99.41	99.64	100.68	98.93		96.98
Developing	100.04	99.65	99.96	101.02	101.92	102.98	103.88	104.63	104.54	104.68		108.59
Centrally Planned	102.06	104.16	106.30	108.50	110.51	112.12	113.47	115.11	116.93	118.86		130.45
Inventories	96.61	91.86	88.78	85.70	83.37	83.21	81.67	81.27	76.74	72.89		63.15
Price	105.28	109.46	111.69	114.61	116.93	117.44	120.57	122.16	129.23	132.93		156.13
Cotton												
Simulation IIa 1 Percent Increase in GDP												

											(1970)
Supply	100.00	100.00	100.01	100.03	100.09	100.21	100.43	100.73	101.05	101.46	98.80
In Developed	100.00	100.00	100.00	100.00	100.00	100.00	100.00	100.00	100.00	100.00	100.00
Developing	100.00	100.00	100.02	100.06	100.12	100.22	100.33	100.43	100.51	100.53	99.00
Centrally Planned	100.00	100.00	100.02	100.02	100.12	100.42	101.02	101.96	103.20	104.60	97.81
Demand	100.38	100.51	100.53	100.46	100.31	100.10	99.88	99.70	99.63	99.73	102.99
In Developed	100.60	100.88	100.96	100.88	100.61	100.17	99.60	98.99	98.45	98.12	106.19
Developing	100.47	100.47	100.46	100.44	100.38	100.26	100.08	99.85	99.60	99.37	101.82
Centrally Planned	100.00	100.00	99.97	99.92	99.87	99.89	100.09	100.53	101.23	102.16	100.77
Inventories	99.34	98.44	97.48	96.61	96.17	96.33	97.28	99.52	102.50	105.56	91.75
Price	100.04	100.25	100.80	101.79	103.17	104.78	106.37	107.56	107.92	107.08	85.25

Simulation IIb 1 Percent Increase in Rate of Growth of GDP

Effect on:

											(1970)
Supply	100.00	100.00	100.01	100.04	100.12	100.34	100.79	101.56	102.65	104.42	115.08
In Developed	100.00	100.00	100.00	100.00	100.00	100.00	100.00	100.00	100.00	100.00	100.00
Developing	100.00	100.00	100.02	100.08	100.20	100.43	100.78	101.26	101.85	102.49	99.54
Centrally Planned	100.00	100.00	100.00	100.02	100.14	100.57	101.61	103.68	107.18	112.53	143.91
Demand	100.38	100.89	101.41	101.85	102.15	102.25	102.08	101.68	101.12	100.65	118.33
In Developed	100.60	101.49	102.46	103.35	103.97	104.11	103.60	102.37	100.44	98.09	116.26
Developing	100.47	100.94	101.41	101.85	102.23	102.50	102.56	102.37	101.88	101.10	103.88
Centrally Planned	100.00	100.00	99.97	99.89	99.75	99.63	99.69	100.19	101.40	103.67	132.96
Inventories	99.34	97.75	95.17	91.61	87.61	83.51	79.64	78.71	84.71	93.12	212.73
Price	100.04	100.30	101.11	102.97	106.37	111.80	119.67	130.14	142.36	153.76	75.10

Sugar

Simulation IIa 1 Percent Increase in GDP

Effect on:

											(1973)
Supply	100.00	100.01	100.16	100.31	100.47	100.63	100.30	99.99	100.13	100.28	100.25
In Developed	100.00	100.04	100.19	100.30	100.42	100.42	100.05	99.92	100.09	100.22	100.16
Developing	100.00	100.00	100.11	100.27	100.42	100.59	100.38	99.98	99.96	100.17	100.19
Centrally Planned	99.99		100.25	100.44	100.64	101.01	100.44	100.11	100.50	100.55	100.48
Demand	100.30	100.30	100.22	100.18	100.21	100.33	100.38	100.33		100.30	100.31
In Developed	100.16	100.21		100.20	100.25	100.30	100.30	100.27	100.26	100.27	100.26
Developing	100.78	100.73	100.69	100.66	100.61	100.75	100.81	100.76	100.71	100.70	100.71
Centrally Planned	99.97	99.86	99.69	99.44	99.59	99.87	99.97	99.89	99.81	99.84	99.82
Inventories	98.88	97.38	96.00	96.14	98.59	100.53	100.16	98.74	97.65	98.01	96.92
Price	101.04	101.75	102.54	103.53	100.71	99.21	100.25	101.45	101.64	100.95	101.27

continued

Table 5-2 continued

	1956	1957	1958	1959	1960	1961	1962	1963	1964	—	1965	1973
Sugar (continued)												
Simulation IIb 1 Percent												
Increase in Rate of												
Growth of GDP												
Effect on:												(1973)
Supply	100.00	100.01	100.18	100.53	101.23	102.43	101.84	100.76	101.18		102.16	104.55
In Developed	100.00	100.04	100.24	100.61	101.34	101.93	100.91	100.31	100.87		101.88	103.43
Developing	100.00	100.00	100.11	100.41	100.98	102.13	102.09	100.81	100.54		101.46	103.62
Centrally Planned	99.99	100.00	100.25	100.74	101.68	103.78	102.51	101.24	102.70		103.77	108.19
Demand	100.30	100.60	100.80	100.79	101.06	101.78	102.35	102.55	102.52		102.80	105.45
In Developed	100.16	100.37	100.57	100.68	101.04	101.50	101.82	102.00	102.12		102.44	104.74
Developing	100.78	101.51	102.20	102.82	103.24	104.38	105.51	106.19	106.69		107.16	114.11
Centrally Planned	99.97	99.80	99.33	98.16	98.17	98.99	99.40	99.03	98.02		97.88	94.96
Inventories	98.88	96.01	90.95	87.12	91.03	96.87	96.01	88.51	79.13		79.94	56.96
Price	101.04	103.02	106.92	115.43	107.96	101.32	104.06	110.94	118.97		113.86	136.62
Wheat												
Simulation IIa 1 Percent												
Increase in GDP												
Effect on:												(1971)
Supply	100.01	100.03	100.06	100.06	100.08	100.12	100.20	100.24	100.76		100.27	100.18
In Developed	100.02	100.09	100.15	100.16	100.17	100.18	100.21	100.19	100.14		100.11	100.13
Developing	100.00	100.00	100.00	100.01	100.04	100.09	100.15	100.20	100.23		100.26	100.13
Centrally Planned	100.00	100.00	100.00	100.00	100.03	100.09	100.21	100.33	100.39		100.44	100.25
Demand	100.21	100.17	100.15	100.13	100.12	100.12	100.13	100.17	100.20		100.23	100.19
In Developed	100.39	100.31	100.28	100.26	100.24	100.23	100.19	100.23	100.26		100.30	100.27
Developing	100.02	99.99	99.94	99.89	99.85	99.82	99.79	99.78	99.79		99.81	99.91
Centrally Planned	100.15	100.15	100.15	100.16	100.17	100.21	100.29	100.36	100.40		100.43	100.31
Inventories	99.01	98.44	98.11	97.83	97.59	96.79	97.27	97.71	98.21		98.56	98.23
Price	100.05	100.25	100.34	100.39	100.43	100.45	100.56	100.46	100.38		100.29	100.35
Simulation IIb 1 Percent												
Increase in Rate of												
Growth of GDP												
Effect on:												
Supply	100.01	100.04	100.10	100.16	100.25	100.37	100.69	100.97	101.13		101.40	102.96

											(1971)
In Developed	100.02	100.11	100.26	100.41	100.58	100.79	101.22	101.42	101.37	101.38	102.48
Developing	100.00	100.00	100.00	100.01	100.04	100.13	100.29	100.49	100.73	101.08	102.13
Centrally Planned	100.00	100.00	100.00	100.00	100.03	100.12	100.33	100.66	101.07	101.56	103.80
Demand	101.21	100.38	100.52	100.65	100.76	100.85	100.89	101.05	101.31	101.61	102.86
In Developed	100.39	100.71	100.99	101.26	101.50	101.70	101.63	102.00	102.33	102.75	104.05
Developing	100.02	100.02	99.96	99.85	99.70	99.51	99.23	98.93	98.68	98.54	98.16
Centrally Planned	100.15	100.31	100.46	100.62	100.79	101.01	101.30	101.66	102.07	102.55	104.93
Inventories	99.01	97.50	95.70	93.50	90.81	84.66	83.50	82.69	82.58	82.27	73.32
Price	100.05	100.30	100.64	101.02	101.47	102.01	103.21	103.44	103.60	103.62	106.75

Rice

Simulation IIa 1 Percent Increase in GDP

Effect on:

											(1971)
Supply	100.02	100.04	100.11	100.20	100.31	100.39	100.44	100.42	100.35	100.22	99.93
In Developed	100.17	100.50	100.98	101.50	101.96	102.24	102.24	101.95	101.42	100.79	100.66
Developing	100.00	100.00	100.00	100.00	100.00	100.00	100.00	100.00	100.00	100.00	100.00
Centrally Planned	100.00	100.00	100.06	100.19	100.39	100.58	100.71	100.73	100.62	100.39	99.68
Demand	100.21	100.18	100.17	100.17	100.18	100.20	100.22	100.24	100.27	100.27	100.06
In Developed	102.08	101.94	101.64	101.16	100.61	100.14	99.90	100.00	100.47	101.22	102.75
Developing	100.03	100.03	100.01	100.00	99.98	99.97	99.98	99.99	100.02	100.04	100.03
Centrally Planned	100.00	100.00	100.05	100.19	100.38	100.57	100.69	100.71	100.61	100.38	99.69
Inventories	98.74	97.89	97.50	97.71	98.63	100.05	101.70	103.13	103.71	103.30	96.22
Price	100.48	101.21	102.02	102.70	103.00	102.78	102.03	100.87	99.64	98.65	101.63

Simulation IIB 1 Percent Increase in Rate of Growth of GDP

Effect on:

Supply	100.02	100.06	100.17	100.37	100.69	101.11	101.58	102.03	102.50	102.72	101.57
In Developed	100.17	100.67	101.68	103.28	105.41	107.93	110.46	112.60	114.00	114.55	113.64
Developing	100.00	100.00	100.00	100.00	100.00	100.00	100.00	100.00	100.00	100.00	100.00
Centrally Planned	100.00	100.00	100.06	100.25	100.65	101.26	102.02	102.82	103.49	103.87	101.83
Demand	100.21	100.38	100.57	100.73	100.88	101.09	101.29	101.54	101.91	102.23	102.45
In Developed	102.08	104.07	105.76	106.96	107.56	107.59	107.34	107.24	107.76	109.26	128.30
In Developing	100.03	100.06	100.07	100.06	100.04	100.01	99.98	99.98	100.00	100.05	100.39
Centrally Planned	100.00	100.00	100.05	100.24	100.63	101.23	101.98	102.76	103.41	103.79	101.79
Inventories	98.74	96.66	94.05	91.56	90.25	90.14	92.17	95.85	100.47	104.29	93.46
Price	100.48	101.69	103.82	106.81	110.21	113.54	115.87	116.57	115.48	113.14	109.68

*multiplier simulation values/base simulation values × 100.0.

simulations show a remarkable picture of price response and dynamic adjustment.

Cocoa

The cocoa model shows substantial price responses to exogenous changes in supply and demand. The price changes evoke several secondary reactions for demand and supply. The latter, particularly, is greatly delayed because of the long time period required to develop new cocoa production.

The price impact of a reduction of cocoa supplies is pronounced— as we see below more pronounced than the response of the coffee market. The 5 percent cut in production evokes an 11.5 percent increase in price in the first year. Price continues to be somewhat above the base solution values for the next two years but falls below the base solution values thereafter. Initially the price decline reflects the lagged response of demand. The supply response does not appear until the seventh year. Note that the effect of a one time change is perceptible even after 10 years due to the long lags in the system. The five percent increase in demand also has a substantial initial effect on price of 15.6 percent. The price increase tends to reduce consumption and ultimately to increase supply, thus eventually decreasing price as above.

Simulation IIa which assumes a continued one percent increase in GDP per capita affects demand though only by a small amount and only in the developing and centrally planned economies. Inventories are reduced and price rises moderately. Ultimately supply increases very roughly in line with demand. The next simulation, which represents a sustained one percent increase in the rate of growth of GDP, accounts for continued expansion of demand. It is not surprising that the expansion of supply lags far behind the growth of demand and that the impact on inventories and on price builds. After 18 years supply almost catches up to demand but at a price some 24 percent above the base solution level.

Coffee

The coffee model multiplier simulation provides a picture of responses that compare well with prior notions of the operation of this market. The reduction in supply of 5 percent evokes an immediate increase in price of 7.4 percent in the initial year of the production decrease. There is an immediate substantial reduction in inventories[1] and almost no demand response in the first year. On the other hand, in the second year demand is reduced by some 2.6 percent but returns to near its base solution level in subsequent years. Price quickly returns to near its base solution level.

In the demand shift simulation, the price effect is somewhat sharper—9.6 percent. Again, as a consequence of the higher price, demand is considerably reduced in year two. The long run adjustment involves a small increase in production, but in view of the long lags involved in the development of additional coffee production, the supply response of some 1.5 percent does not occur until eight years later. A further small price reduction occurs at that point as a result of the increased supply of coffee. Simulation IIa follows closely the pattern we observe above. The impact of a 1 percent increase in GDP on coffee demand is quite small in the developed countries and nonexistent in the developing countries where the negative impact of price reduces consumption. On the other hand, the growth of GDP has a proportionately larger impact in the centrally planned block, where consumption is very small, however. As a consequence, the impact on world demand is very small, 0.16 percent in the first year and a little less as a result of the price effects in subsequent years. Nevertheless, since the impact on supply is so long delayed and so small, there is a perceptible impact on price. The coffee price increases relative to the base solution by 0.2 to 0.3 percent over most of the simulation period, with somewhat greater price increases observed late in the simulation as continued small declines in inventory levels, compared to the base solution, are apparent.

The continued more rapid growth of GDP in simulation IIb produces a more perceptible buildup of world demand. Again, the impact is on inventories since increases in supply are long delayed. Price increases noticeably by between 2 and 2.5 percent after ten years relative to the base solution. Late in the simulation period we note a somewhat greater buildup of price by 9 percent relative to the base solution in the eighteenth year. At this point the price is rising sufficiently to prevent further increases in demand or compared to the base solution, despite the continued increase of GDP at a rate 1 percent per year faster than in the base solution. Production has also increased at this point by 0.7 percent relative to the base simulation, almost as much as demand but not yet in line with demand so that inventories are still declining and further price increases may be anticipated.

Tea

The somewhat slow adjustment of the tea model to short run disturbances noted above produces a somewhat delayed response to the reduction in production assumed in simulation Ia. The immediate impact of the reduction in output, which is analogous to a substantial incident of poor weather conditions, is in large part on inventories with only minimal reductions in consumption. The price increases by 1.1 percent relative to the base simulation in the first year but then continues to increase to a level 2.2 percent above the base solution in year four. The higher price evokes a small supply response in the developing countries, and eventually this increase in supply and the price-induced cut in demand cause the model to return close to its base simulation values. In the long run price levels are about 1 percent less than in the base simulation.

The impact of a 5 percent increase in demand is analogous. The initial increase in demand affects price more substantially—an increase of 4.8 percent relative to the base solution in the first year. Consequently subsequent adjustments of demand are greater and the increase in supply, particularly when the effect of new planting becomes apparent some six to eight years later, is more perceptible—an increase in output of 1 percent over the base simulation. Ultimately the model settles down to values which are very similar to those observed with a 5 percent disturbance on the supply side.

In view of the small impact of GDP on demand in the case of tea, the impact of simulation IIa is not very perceptible. GDP affects consumption of tea only in the developing countries, which account for a relatively large share of consumption. The increase in GDP increases tea demand worldwide by only 0.1 percent. This produces a price increase of only 0.1 in the first year, but the price increase relative to the base solution builds up to 0.5 percent after seven or eight years as the higher level of GDP is maintained. This evokes a gradually growing response from the supply side so that the impact on supply exceeds the impact on demand after seven or eight years. Ultimately the price adjusts to a level very close to that of the base simulation.

The continued faster growth of GDP postulated in simulation IIb extends the results of the case above. The persistent growth of demand relative to the base solution in the developing countries is offset partially by reduced consumption in the developed and the centrally planned economies as a result of the impact of higher tea price. Price tends to rise gradually throughout the simulation period and eventually evokes increases in supply sufficient to offset the higher level of demand. At the end of the simulation the impact on supply greatly exceeds the impact on demand and inventories are returning close to the base solution level. The price is continuing to show some increase at a level some 7 percent above the base simulation.

Wool

The model for the wool market displays important immediate response properties and considerable variation from year to year. A 5 percent cut in supply causes an almost corresponding reduction in inventory and a substantial upward response of 8.5 percent in price. While there is little contemporaneous adjustment of demand, in the following year demand is cut sharply, particularly in the developed market economies. The reduced price evokes a continued cutback of output in the second year of the multiplier simulation but after that the market adjusts back to normal quickly. Several years after the multiplier disturbance on the supply side, only insignificant fluctuations remain and the multiplier solution is very close to the base solution values for all variables. The postulated 5 percent increase in demand brings about an immediate increase of 16 percent in price and an immediate supply adjustment amounting to 1.3 percent, all of which occurs in the developed wool-producing countries. The striking consequence is that wool consumption responds with a substantial reduction in the following year of about 4 percent below the base solution level. This cut in consumption is accompanied by a moderate reduction in production and decline in price to 5.3 percent below the base solution. In subsequent years we observe some lagged supply response to the initial higher price from the developing mixed and the centrally planned economies. Fluctuations in demand, production, price, and inventories gradually fade away so that by the end of a ten year period all variables are again very close to their base solution values.

The 1 percent increase in GDP in simulation IIa translates into an immediate increase in wool consumption of 1.7 percent, with most of the change in the developed market and the centrally planned economies. The initial pattern of movement is similar to that produced in simulation Ib, with a moderate increase in output, an increase in price of 5.3 percent, and a tendency toward cycling reactions. Ultimately, however, the market settles down to consumption approximately 0.6 to 0.7 percent above the base solution and a corresponding increase in production. The price stabilizes 2.0 to 2.5 percent above its base solution value and inventories tend to remain about 2 percent below the base solution. Simulation IIb imposes a much more severe strain on the world's wool industry. While the initial adjustment is of course the same as in simulation IIa, the continued growth of demand in response to the more rapid growth of GDP causes a continual downward adjustment of inventories and upward movement of price. Supply responds with a lag as more production becomes available largely from the developing countries and from the centrally planned block. Cyclical variations are imposed on a continued upward movement of demand, price, and supply. Toward the end of the simulation period, supply is almost as much above the base period simulation, about 7 percent, as demand, suggesting that supply and demand may be returning to balance. At this point, however, inventories are more than 30 percent below the base solution inventories and the price is over 50 percent greater than in the base solution. Because of the variation at this time, we cannot be sure that balance between demand and supply will be maintained and that the upward movement of price relative to the base solution will cease.

Cotton

The cotton market multiplier simulations reveal some curious properties of the cotton model. While the initial responses of the model appear reasonable if somewhat slow, later in the simulation the initial disturbance results in the gradual buildup of cyclical movements which results in a breakdown of the solution. The source of this problem may lie in the fact that the production of the developed market economies—which consists principally of production in the United States—has been isolated from the rest of the model through the use of United States acreage allocations and the United States cotton price as explanatory variables. The price variable was substantially separate from world market price over the simulation period as a result of U.S. agricultural policy. Moreover, the world market price is heavily dependent on its own lagged value, a matter of concern as we have indicated above.

The initial response to a reduction of 5 percent in production is only a very small increase in price, 0.4 percent, and no change in demand. Thus the cut in production is entirely absorbed by reduction in inventories. In the subsequent years we see a gradual buildup of price increases to 13 percent above the base solution in year six, which results in moderate increases in production, particularly in the centrally planned economies block, after some years' lag. The price increases also provoke gradual downward adjustment of consumption of approximately 5 percent by year six. Stocks are gradually rebuilt to above their base solution level until year ten, following which the model begins to swing violently until simulated inventories fall below the zero level. This is clearly an unacceptable result but one that reflects the very gradual buildup of the unstable cyclical property of this model system.

The demand stimulus simulation Ib produces results which are very similar. As above, the buildup of price gradually reaches 12 percent in year nine; the response of supply is delayed but eventually reaches 3 percent above the base solution. Late in the simulation period the unstable cyclic property of the model asserts itself in this simulation as above.

The 1 percent shock to GDP in simulation IIa represents a less serious disturbance to the model and permits solution for the entire simulation period. The initial impact of higher GDP produces an increase of 0.4 percent in demand but only one-tenth as much, 0.04 percent, in price. Since there is not initially any adjustment in supply the entire change is absorbed by inventories that drop gradually over the first five years of the simulation period to approximately 4 percent below the base solution. Over the same period price gradually builds up to some 8 percent above the base solution; output increases after some time lag, particularly in the centrally planned economies. Production reaches a point approximately 2 percent above the base solution in years eleven and twelve. The impact of higher cotton price gradually reduces demand below the base solution levels. Over time we observe a gradual buildup of inventories above base solution levels, and toward the end of the simulation period this provokes a sharp decline of prices, evidence of the beginning of the unstable cyclical behavior noted in the simulations above.

In simulation IIb, as above, the initial impact of more rapid growth of GDP is on inventories and price. There is no supply response at all from the developed countries, since this supply depends on variables which are not related to the world cotton economy as we have noted above. This may be reasonable in the short run but represents a serious handicap from the standpoint of longer term dynamic simulations of this model. After some time lag, supply in the developing countries and in the centrally planned block responds to higher price. At the same time, the high price of cotton tends to limit the growth of demand, even though GDP is increasing consistently more rapidly than in the base solution. Thus, after five years demand is up approximately 2 percent, price is up 6 percent, and supply is up only 0.3 percent. But in subsequent years the strong lag coefficient of the price equation begins to push price upward systematically, resulting in some decline in demand, and ultimately in increasing supply particularly from the centrally planned economies. Eventually inventories increase and price drops, with the wide cyclical swings which ultimately appear to be typical of this model system.

Sugar

The responses to a 5 percent decrease in supply (simulation Ia) correspond closely to what would happen if there is a one time widespread shortfall of the sugar harvest. The consequence of reduced production is to drive up current price sharply by almost 3 percent for every 1 percent shortfall in output. There is a very small immediate effect on demand. In the next periods, supply not only comes back (only a one time disturbance has been assumed) but rises somewhat above its base solution level in response to the higher price. The lag response creates a diminishing cycle for price which falls somewhat below the base year level before finally adjusting back very close to its base solution path. The simulation with a corresponding 5 percent one time stimulus to demand (simulation Ib) shows a similar pattern of reaction. A small part of the exogenous 5 percent stimulus to demand is immediately offset by the effects of higher price. The impact on price is a little more pronounced in simulation Ib than in simulation Ia since the demand enters directly into the price determination equation.

The stimulus to GDP simulations IIa and IIb are intended to show the sensitivity of the sugar market to the expansion of world economies. As one would expect a 1 percent increase in GDP above base solution levels (simulation IIa) has only a small impact on demand, since the income elasticities are small, particularly in the developed and centrally planned economies. The effect on price is quite perceptible, however, amounting to 1 percent in the first year and rising to 3 percent after three years. At this point the supply response becomes important, and price is driven down. There is some evidence of cycling though with limited amplitude. The price tends to fluctuate within a range of 0.5 to 1.5 percent above the base solution level, and the supply response tends to be approximately in accord with the expansion of demand.

In simulation IIb the impact of faster income growth is pronounced. While the growing price stimulates faster growth of supply, increased production never quite catches up with the increasing demand, so that rapid and substantially continuous increase in the price of sugar results. A feature of this simulation is that stocks tend gradually to decline as the demand continuously exceeds the supply available from current production. This is an important finding since it suggests that more rapid growth, particularly in the developing economies where the income elasticity for sugar is still relatively high, would tend to increase the price of sugar.

Wheat

The behavior of the wheat model comes very close to prior expectations on the basis of the single equation structure of the system discussed above. The 5 percent decrease in supply postulated in simulation Ia has been assumed across the board in all producing regions. This is thus a case similar to a worldwide drought situation, a case which is somewhat different from a regional bad harvest situation such as that which occurred in 1972–1973. Since consumption in both the developed countries and in the centrally planned blocks is directly linked to production in these areas—though probably as a result of inventory measurement problems in the former and as a consequence of deliberate economic policy in the latter—there is a significant adjustment of demand to the cut in output. World demand is reduced by approximately 2 percent in response to a 5 percent drop in output. This provokes only an 0.5 percent increase in the price in the first year—a result which may reflect the inability of the model to catch anticipations phenomena. This increase appears particularly small in view of the fact that reduced output in the centrally planned economies accounts for a 32 percent increase in their imports.[2] In subsequent years we observe a response on the supply side, but at first only in the developed countries. This response amounts to 0.3 percent in the year following the supply reduction but gradually mounts to 0.8 percent as supplies are increased from the centrally planned economies. Ultimately the model returns to the base simulation path, so that after ten or more years the principal variables are back where they would have been without a supply side disturbance.

A substantial change on the demand side is a less realistic test of the model's performance since it is unlikely that substantial exogenous variations in demand would occur. Again it has been assumed that the change occurs in all consuming areas. The impact is not off-set here by an adjustment in supply since production has not been altered and only very small current production responses can occur. Thus the principal impact of higher demand is a significant reduction in inventories and an increase in imports of the centrally planned countries. Inventories are reduced 22 percent relative to their base solution levels and imports of the centrally planned countries rise 52 percent. In view of these substantial changes it is surprising that the price impact in the initial year is no larger than 1.7 percent. It rises further to almost 5 percent in the second year but these changes doubtless understate the true impact of changes in inventories and in imports by centrally planned countries on wheat prices. In subsequent years we see as before some upward adjustment of supply and eventually the model returns very close to its base solution path.

Demand for wheat is sensitive to the GDP both in the developed and the centrally planned economies. An increase in GDP of 1 percent, assumed in simulation IIa, raises world demand by approximately 0.2 percent, though the resultant increase in the price of wheat tends subsequently to reduce consumption in later years. As we have noted earlier the price response is somewhat lagged so that price increases to a level 0.5 percent above the base solution after a delay of six years. The supply response is also lagged, particularly in the developing and the centrally planned economies. Ultimately supply tends to increase roughly in accord with the increase in demand so that the model tends to adjust to a level of GDP higher than the base solution. The equilibrium price appears to be approximately 0.2 percent higher than in the base solution.

The continued more rapid growth of GDP assumed in simulation IIb causes continued growth of demand relative to the base solution over the simulation period. After the anticipated delay due to the slow adjustment of output in the developing and centrally planned economies, output increases approximately in accord with the increase of demand at levels almost 3 percent above the base solution after 15 years of simulation. At this time the price also appears to stabilize at a level between 3 and 4 percent higher than the base solution. In the final year of the simulation, 1971, an aberration introduces a new fluctuation and drives the model from the equilibrium path.

Rice

The rice model reflects, more than the other commodity models, the fact that this product is an essential staple which is only marginally in world trade. Much rice is consumed where it is produced and consumption relates directly to production in the developing countries and in the centrally planned economies. Thus a cutback in production, as is assumed in simulation Ia, has its immediate impact by reducing consumption by an almost corresponding amount. Moreover, since there are lag effects in the supply functions for the developed and the developing economies, supply cutbacks are continued in succeeding years. World demand is held down not only by the lower production but also by a response to price that rises gradually to a level 6 percent above the base simulation in year four, as a consequence of reduced inventories. There is a little more supply response in this system than appears on the surface in this simulation since some of the downward impact of the initial reduction in output is offset by response to higher price, initially in the developed countries and later in the centrally planned block. As we have noted above there is no price impact in the equation for supply for the developing economies. The delayed supply response causes fluctuations in price, demand, and inventories which have not faded away by the end of the simulation period.

The comparable change on the demand side, simulation Ib, has considerably more impact since there is not the corresponding adjustment of supply which we noted above. As a result, the price of rice is driven immediately to a level 13 percent above the base solution and rises further to 21 percent above the base solution by year three. Supply responds quickly to higher prices only in the developed countries and then later with some lag in the centrally planned economies. There is thus an immediate downward impact on inventories of almost 30 percent, but when demand begins to fall back to its original level and responds to the higher price of rice demand falls below supply and inventories rebuild. Inventories rise substantially above their base solution level with consequent downward impact on price. The result of the disturbance on the demand side is to produce wide cycles that may be of as much as seven years' duration. It is not possible to tell on the basis of the simulation runs whether these cycles would eventually die away or whether they would expand further.

The 1 percent increase in GDP in simulation IIa has an impact principally in the developed countries, where it provokes a 2 percent increase in demand, but since these countries account for only a small proportion of world consumption the impact worldwide is only 0.2 percent. The production response is slow, and price of rice increases in the first year to approximately 0.5 percent and in the next two to four years to almost 3 percent above the base solution level. The higher price eventually evokes higher production by up to 0.4 percent—somewhat in excess of the increase in demand. This causes inventories to rise above the base solution level, and we observe a downward movement of price. At the close of the simulation period fluctuations are still observable, though the principal variables are once again very close to their base solution levels.

As we would anticipate, the impact of higher growth of GDP in simulation IIb is considerably greater. The buildup of demand in the developed countries, and somewhat later in the centrally planned economies, as a result of the faster growth of GDP is greatly influenced by the pattern of price which rises gradually to 16 percent above the base solution level after eight years. Supply increases with a lag, particularly in the developed countries. There is no supply response in the developing countries, an aspect of the model which may be questionable given the long term and persistent character of the expansion of GDP assumed here. Ultimately supply builds up into line with demand which is held in check by higher price despite the continued more rapid growth of GDP assumed here. At this point we observe some decline in price to fluctuations between 5 and 10 percent above the base solution level.

CONCLUDING COMMENTS ON
MULTIPLIER SIMULATIONS

The multiplier simulations of the eight commodity market models demonstrate consistently the diverse responsiveness of the commodity markets to external stimuli. While the markets differ significantly, in all cases we were able to observe dynamic responses, frequently in accord with our prior expectations on the nature of the commodity and its market. In many instances the markets tended to readjust as expected to the exogenous change, though in one case—cotton—cyclical instabilities gradually developed from which the model solution did not recover. It is not clear whether in all cases the response of the models to disturbances are sufficiently large and sufficiently quick. This was particularly apparent in the case of wheat.

The principal conclusions which can be drawn from the multiplier simulations may be summarized as follows:

1. A one time disturbance causes reactions in the models which in almost all cases cause variables ultimately to return to the base simu-

lation path. This tendency for readjustment, which operates through the demand and the supply sides, was apparent except, as noted above, in the case of the cotton model.

2. The initial impact of demand changes on price is greater than the effect of exogenous changes of supply, though the latter are likely to be greater in amplitude in the real world.

3. Responses of price to exogenous changes vary considerably. Price volatility is particularly great in the cases of coffee, cocoa, and wool. Price adjustments in wheat, rice, and cotton are more moderate and the lag adjustments found may be unrealistically slow. As we note earlier, it is probably important to capture expectational phenomena in these models.

4. Demand side responses to higher prices are delayed, but of particular importance is the fact that the response on the supply side is slower still, as we have noted above. The responsiveness of supply depends greatly on the commodity—tree crops of course respond more slowly than field crops—but depends also on the producing region. The slowness of the response on the supply side is an important source of cyclical movement of the models, though we have noted that the cycles tend to fade away except in the case of cotton.

5. The slowness of the adjustment of supply means that it takes considerable time for the model to adjust to persistent changes in demand. But such adjustment was observed in most cases. However, acceleration in the rate of growth of GDP tended to provoke continued increase in price since the required adaptation of supply was not always achieved in the simulation period analyzed here.

NOTES

1. Since the level of the inventory numbers is rather arbitrary, the effect on inventories in percentage terms is only intended to give an approximate indication.

2. The willingness of the USSR to meet domestic crop shortfalls by importing wheat has increased greatly in recent years. This means that world markets now are considerably more affected by crop conditions in the centrally planned countries than during the sample period considered here.

Chapter 6

Summary and Conclusions

This study has produced operational econometric models for eight agricultural commodities. These models, simple yet structural in specification, have been described in full detail above. The research yields numerous general conclusions which are summarized in this chapter. To organize them, they are divided into five groups—supply, demand, price and inventories, sample period dynamic simulations, and multiplier response simulations. Concluding remarks at the end of this chapter reflect on our successful experience with these models and comment on their promise as a basis for future development and for applications.

SUPPLY

Table 6-1 presents important characteristics of supply by commodities and country groups. The price elasticities that were estimated in the equations presented above have a number of interesting features.

1. Supply price responses are pervasive. Only for rice production in the developing economies is there no evidence of a significantly nonzero response.

2. Aggregation of the price elasticities by average shares to obtain world supply price elasticities for each commodity leads to three groups of commodities—wheat and cotton with long run world price elasticities in the range 0.71 to 0.75; tea, coffee, cocoa, and sugar with elasticities in the range 0.30 to 0.34; and rice and wool with

Table 6-1.   Major Characteristics of Supply of Eight Commodities[a]

Characteristics and Country Groups	Commodities							
	Cocoa	Coffee	Tea	Wool	Cotton	Sugar	Wheat	Rice
Long Run Price Elasticity of Supply								
Developed	0.34	0.33	0.09	0.04	1.34	0.15	0.38	1.54
Developing			0.23	0.54	0.07	0.19	0.57	0.00
Centrally Planned			0.72	0.61	1.02	0.71	1.03	0.25
Mean Lag in Price Response (years)								
Developed			4.0	1.0	1.0	1.7	0	6.5
Developing	8.0	8.4	7.0	4.0	1.0	2.5	4.5	
Centrally Planned			5.1	3.0	4.3	2.7	5.0	2.6
Secular Trend (percent per year)								
Developed			1.8	1.4	6.3	4.2	4.0	0
Developing	4.9	2.4	4.0	4.3	3.7	3.9	5.3	2.8
Centrally Planned			5.6	1.3	7.7	6.3	6.4	2.6

[a]The data and sources are given in Appendix B. Chapter III presents the supply equations.

elasticities in the range 0.21 to 0.25. This ordering primarily reflects the ease of shifting factors in and out of these commodities in the long run. Much of rice production in the developing and centrally planned economies, for example, is on land with irrigation systems that do not permit the control and timing needed for alternative crops. In addition to the question of the ease of substituting factors in explaining the pattern of elasticities, however, there is also the impact of policies which modify the relationship between foreign and domestic prices.

3. The world price elasticities of supply for these commodities suggest that such responses are not only significantly nonzero, but fairly substantial in comparison with other estimates (e.g., Nerlove [1958]).

4. For the six commodities produced in each of the three country groups some comparisons can be made across the type of economies. The equally weighted long run price elasticities of supply are 0.72 for the centrally planned economies, 0.59 for the developed economies, and 0.27 for the developing economies. The estimates imply that the centrally planned economies probably are the most price responsive. This surprising result does not mean that in such economies prices are used extensively for internal allocation, but that governments are quite conscious of long run opportunity costs as represented by relative international market prices when they make allocation decisions. The estimates imply that the developed economies are almost as price responsive, because the very high responses for rice and cotton offset substantially the relatively low ones for tea, wool, sugar, and wheat. For each of these last four commodities, the estimated elasticities are higher for the developing than for the developed countries, in fact, but the negligible price response for cotton and rice in the developing economies leads to an overall average for that group substantially below those for the other two country groups. The imposition of substantial barriers between international and domestic prices in the developing economies for both of these crops is one factor which underlies these low estimated elasticities. Another consideration for rice—which is relevant for understanding the low rice price elasticity in the centrally planned group in comparison to the other responses within that group—is that much of the production never enters the market but is consumed instead by the producing units.

The second part of Table 6-1 gives the mean lag in years for the price responses. For obvious biological reasons related to the gestation period between planting and bearing, the longest mean lags are for the tree and bush crops of coffee, cocoa, and tea. The pattern for the other five commodities, however, reflect institutional considerations much more than biological ones. Because the developed, mixed capitalist economies are most integrated into the world economy, this country group has the smallest mean lag for five of the six commodities produced in all three regions. The lags for the other two country groups are much longer, in part because the internal situation is more isolated from fluctuations in the international market, especially for such staples as wheat. This may reflect conscious efforts by government regulation to protect such staples as wheat or it may simply follow from the limited participation in the market economy of producers in less developed countries. For the centrally planned economies, moreover, more time is required for the bureaucracy to function and to be sure that relative price changes do represent longer run opportunity costs and not just short run fluctuations.

The third part of Table 6–1 refers to the underlying secular trends around which the price responses occur. Generally these are higher for wheat, cotton, and sugar than for the other commodities and higher in the centrally planned group than in the developing economies. This pattern across country groups reinforces the observation made above in the discussion of shifts in relative shares over the sample period about the relative dynamism across these groups. Note, however, that the reduction in shares for the developed economies is not due only to lower underlying secular trends. For example, for cotton, for which the developed group lost the largest share, the estimated secular trend was relatively high. In this case the reduction in the developing countries' share primarily reflected the relatively large price response therein to the secular declining real price.

DEMAND

Table 6–2 presents major characteristics of demand for the eight commodities. The long run price elasticities of demand for the eight commodities by the three country groups obtained from the demand equations above, have several interesting features.

1. Price responses, once again, are pervasive. Only for the centrally planned economies in the cases of the staple crops is there no evidence of significantly nonzero price responses in demand. On the average—

Table 6-2. Major Characteristics of Demand for Eight Commodities[a]

Characteristics and Country Group	Commodities							
	Cocoa	Coffee	Tea	Wool	Cotton	Sugar	Wheat	Rice
Long Run Price Elasticity of Demand								
Developed	-0.33	-0.24	-0.07	-0.27	-0.44	-0.03	-0.52	-0.8
Developing	-0.13	-0.31	-0.14	-0.16	-0.18	-0.05	-0.51	-0.0
Centrally Planned	-0.63	-1.25	-0.48	-0.20	-0.14	-0.47	0.00	0.0
Mean Lag in Price Response (years)								
Developed	1.3	1.0	0	3.8	1.9	0.3	0	2.2
Developing	3.1	1.5	2.0	0.6	2.8	1.0	1.5	1.2
Centrally Planned	1.6	0.5	1.6	5.2	0.3	2.5		
Long Run Income or GDP Elasticity								
Developed	0	0.20	0.00	1.81	1.15	0.30	0.42	2.1
Developing	1.38	0.40	0.43	1.05	0.47	0.78	0.03	0.0
Centrally Planned	1.18	1.45	0.00	2.05	0.00	0.36	0.16	
Mean Lag in Income of GDP Response (years)								
Developed	3.1	0	0	0	0.09	0.3	0	0
Developing		1.0		0	0	0.1	0	0
Centrally Planned	0	0		0		2.5	0	
Secular Trend (percent per annum)								
Developed			-0.05	-6.4	-5.1		-3.1	-11.0
Developing				-9.6	0.3			
Centrally Planned								
Production Elasticity								
Developed							0.29	
Developing								0.9
Centrally Planned					0.75		0.63	0.9

[a] Appendix B gives data and sources. Chapter III presents the estimated demand functions.

although not for every commodity or country group—the absolute magnitudes of the estimated long run price elasticities of demand tend to be smaller than those for supply.

2. Aggregation of the price elasticities by average shares to obtain world price elasticities for each commodity leads to some homogeneity. The estimates of –0.07 for rice, –0.05 for sugar, and –0.15 for tea fall below the –0.24 to 0.32 range of the other commodities.

3. Examination of the price elasticities across commodities for each of the three country groups leads to the conclusion again that the developing economies are less price responsive than are the other two groups. The equally weighted long run price elasticities of demand are –0.34 for the developed economies, –0.40 for the centrally planned economies, and –0.19 for the developing ones. The relatively great degree of price response on the demand side as well as on the supply side for the nonstaple products by the centrally planned economies merits mention. Explanations for these patterns across country groups would seem to be parallel to those offered above for the supply side.

The second part of Table 6–2 gives the mean lag in years for the price responses. A comparison with Table 6–1 reveals that generally the demand lags are less than the supply lags. The biological and technological constraints on shifting supply factors and the formation of expectations on the supply side thus seem to require longer periods for price adjustment than the technological constraints and expectation formation processes on the demand side. The two fiber crops, however, provide exceptions to this generalization for two of the three cells. The relatively long lags in these cases may be due to the relatively high fixed costs and changes in technology involved switching to synthetic alternatives.

The most striking feature which examination of the lags across country groups reveals is the relatively slow adjustment in the developing countries. For five of the eight commodities, in fact, the longest lag is for this group. The fourth part of Table 6–2 indicates, moreover, that three of the lagged demand responses to income are in the developing countries. The inability to adjust quickly to changes on either the demand or the supply side thus seems to be associated with a state of lesser economic development.

The third part of Table 6–2 includes the long run per capita income or GDP elasticities. Aggregation across country groups with relative shares as weights leads to three subdivisions among the seven commodities—wool, with a world elasticity of 1.78; cotton, cocoa, and sugar with world elasticities, respectively, of 0.48, 0.40, and 0.55; and the other four commodities with world elasticities in the range of 0.15 to 0.28. If an elastic income response is used to define a luxury, therefore, only wool so qualifies on the world level (although cocoa, coffee, cotton, and rice qualify for at least one country group).

Aggregation across commodities for each of the country groups leads to an average of 0.75 for the developed countries, 0.57 for the developing countries, and 0.65 for the centrally planned economies. These averages imply that the declining demand shares of the developing countries do not originate in low income elasticities due to satiation (at least for the nonbeverage products). They also indicate that the most limited income response—as for the supply and demand price responses—tends to be in the developing group. Finally, they suggest that the income elasticities on the average tend to increase with per capita income, even though for some commodities (e.g., cocoa in the short run, coffee, and wool) there is evidence of an "S shaped" response with the middle income countries having the greatest income response, and in other cases (e.g., cocoa in the long run and tea) the highest elasticities are for the lowest per capita income country group.

The fifth part of Table 6–2 gives the estimated secular trends. With the single exception of the small positive one for cotton demand in the centrally planned economies, all of the significantly nonzero trends are negative. Presumably they reflect the increasing availability of synthetic substitutes and shifts in tastes away from the commodities being studied. One of these, a large –9.6 percent per year, is for wool demand in the centrally planned economies. For the nonbeverage commodities the shift in demand shares away from the developed countries is associated primarily with these negative secular movements.

The sixth part of Table 6–2 gives the elasticities of per capita demand with respect to per capita production within the respective country groups. This variable is included in some of the demand relations because of the autarkic tendency of some country groups in regard to the source of supply of basic agricultural commodities.[1] Significantly nonzero and substantial elasticities are reported for rice in the developing countries and for cotton, wheat, and rice in the centrally planned economies. For such commodities and country groups, the rationale for including this variable seems to be reasonably persuasive. The smaller elasticity estimated for wheat in the developed economies, however, seems more open to question.

PRICE-INVENTORY

The estimated price relations can be rationalized, alternatively, as inventory demand relations normalized on the deflated prices (see Chapter Two).

Table 6–3 includes some of the major features of these relations. The first part of this table gives the long run elasticities of deflated prices with respect to world inventory to world demand ratios. The range of these elasticities is quite large—from –0.16 for wheat to –12.0 for tea. They imply quite substantial variance in the extent to which a standardized random shock is absorbed by price changes as opposed to inventory changes. In the long run the eight commodities can be ordered as follows in terms of the relative absorption of such shocks by prices instead of inventories—tea, sugar, cotton, rice, cocoa, wool, coffee, and wheat. As the second part of this same table indicates, however, the estimated speeds of adjustment for tea, cotton, and rice are very slow. In the short run, changes in inventories for these three goods absorb much more of any given disturbance relative to prices than in the long run. Sugar, however, shows sharp price responses quickly. Except for sugar, the range of short run elasticities, although considerable, is much less than that for the long run.

The third part of Table 6–3 gives the significantly nonzero secular trends. For no commodities was a positive trend estimate obtained, but negative ones were estimated for four—wool, wheat, cocoa, and coffee. Reasons for such secular shifts include reduced desired inventories due to economies of scale and reduced desired inventories due to improved transportation and communication. That no such trend is included in the table for the three commodities that have very slow price adjustment to the stock to demand ratio may appear to be more than coincidental. Prima facie it might appear that the lagged price terms in the relations for these three commodities are representing the secular trend since the deflated prices for most of these commodities tended to decline over the sample. Exploration of this possibility, however, did not uncover much support for it.

Two shortcomings of the price-inventory relations used in this study should be mentioned. First, there is some evidence that the

Table 6-3. Major Characteristics of Price-Inventory Relations for Seven Commodities[a]

Characteristics and Country Groups	Commodities						
	Cocoa	*Coffee*	*Tea*	*Wool*	*Cotton*	*Sugar*	*Wheat*	*Rice*
Long Run Elasticity of Price with Respect to Inventory/Demand Ratio	-0.86	-0.39	-12.0	-0.66[b]	-5.86	-6.6/9.1[c]	-0.16	-4.2
Mean Lag in Response of Price to Inventory/Demand (years)	0	0	9.6	0	12.9	0[d]	1	11.3
Secular Trend (percent per year)	-2.6	-1.6	0.0	-9.0	0.0	0.0	-4.0	0

[a]Chapter III gives the price-inventory determination relations.
[b]For wool the elasticity of price with respect to world demand is 1.70.
[c]A nonlinear relation has been used which yields an elasticity of approximately -6.60 when inventories are highly relative to demand and -9.1 when inventories are low.
[d]There is additionally a small stock change effect.

relation between prices and inventory to demand level is not log-linear when that ratio becomes very low. Instead, there seems to be an intensified impact on prices. For some commodities this effect is treated in the study through the use of dummy variables (e.g., rice). In the case of sugar a nonlinear relation has been used. A similar method of introducing this nonlinearity into the functional form probably would be preferable in other cases as well.

Second, price expectations are treated only by distributed lag representations and costs of holding inventories are not represented explicitly. In future work the incorporation of such factors might be valuable.

SAMPLE PERIOD COMPLETE MODEL
BASE DYNAMIC SIMULATIONS

The relations discussed above are integrated for each commodity to form world market models. One of the first tests of these models was to examine their performance over the sample period. Dynamic simulations—i.e., in which lagged simulated values of endogenous variables are used instead of actual values—over at least 15 years were conducted. Table 6–4 summarizes the resulting average absolute percentage discrepancies between the actual and simulated values for the three major aggregates in each model.

This table suggests that the models track the world supplies quite well. For all eight commodities the mean percentage errors are less than 5 percent, and for rice, wool, and tea they are less than 2 percent. The general supply movements are captured well, although some fluctuations due to special weather conditions are not replicated so well. That the models perform most poorly on the supply side for the tree crops suggests that their long gestation periods may introduce special problems.

The demand movements are traced even better than those for supply. For every commodity except wool the mean percentage error is lower for world demand than for world supply. In every case the average percentage deviation is less than 3 percent.

The price movements are not traced as well as the quantity movements. Nevertheless, with the exception of cocoa and sugar, the simulations for the deflated prices seem quite satisfactory. The mean percentage deviations range from 3.2 percent for rice to 7.5 percent for wool, with the exception of 13.1 percent for cocoa and 11.3 per-

Table 6–4. Average Absolute Percentage Discrepancies Between Actual and Simulation Values for Major Aggregates in Dynamic Simulations of 15 Years or More[a]

Variables	Commodities (percent)							
	Cocoa	Coffee	Tea	Wool	Cotton	Sugar	Wheat	Rice
World Supply	4.2	4.4	1.9	1.3	3.2	2.4	2.5	1.2
World Demand	2.6	2.6	1.4	2.5	2.6	1.1	2.0	1.0
Deflated Price	13.1	4.4	6.4	7.5	4.1	11.3	3.4	3.2

[a] Appendix C gives the simulation results.

cent for sugar. Given the price fluctuations in some of these markets, such deviations are not unreasonable. Although some minor fluctuations are missed, the general movements are captured fairly well. There seems to be no obvious association, incidentally, between success on the individual commodity level and the sizes of the elasticities or the lengths of adjustment included in Table 6–3.

DYNAMIC MULTIPLIER SIMULATIONS

Multiplier simulations were carried out to test the performance properties of the commodity market model systems. In each case comparisons were made between a so-called "disturbed" simulation and the corresponding base simulation, the sample period dynamic simulations discussed above. Multiplier simulations were calculated for exogenous changes in supply and demand, in each case a 5 percent one period exogenous change was assumed. Multiplier simulations were also calculated to measure the impact of sustained changes in GDP, a sustained 1 percent increase above the base solution in one alternative, and a sustained 1 percent increase in the rate of growth of GDP in the other. Detailed discussions of the multiplier simulation results are provided above.

In Table 6–5 we summarize the impact on price of the exogenous changes in demand and supply. In order to give an indication of short run and long run effects, the table shows the impact in the year of the exogenous disturbance (year one), the year following (year two), and in the ninth year following (year ten). In each case the figures indicate the level of the price in the multiplier solution relative to the level of price in the base solution multiplied by 100.

It is apparent that all the commodity prices respond to changes in supply and demand. The demand side changes have consistently greater impact on price, though we should note that in the real world fluctuations on the supply side as a result of weather are likely to have greater amplitude for these agricultural commodities than fluctuations on the demand side.

Table 6–6 shows similar statistics for the simulations assuming persistent exogenous changes in GDP. The effect of changes in GDP varies greatly among the commodities, partly as a result of very different sensitivity of demand to GDP and partly as a result of different model response properties. Increases in price result in all cases, though these are small except in the case of wool (we should remem-

Table 6–5.   Effect of 5 Percent Exogenous Changes in Supply and Demand on Price*

	Cocoa	Coffee	Tea	Wool	Cotton	Sugar	Wheat	Rice
5 Percent Decrease in Supply								
Year 1	112.1	107.8	101.1	108.5	100.3	114.8	101.1	100.4
Year 2	104.8	101.4	101.9	94.0	101.4	106.4	102.2	102.4
Year 10	95.8	99.6	100.0	99.6	95.0	99.7	99.5	93.0
5 Percent Increase in Demand								
Year 1	115.6	109.4	104.8	115.5	100.6	117.1	101.7	113.4
Year 2	102.5	100.4	104.3	95.0	102.7	106.0	105.0	120.7
Year 10	95.7	99.4	98.5	99.7	88.4	99.6	99.3	78.3

*multiplier simulation value/base simulation value × 100.0

ber that the exogenous change in GDP is only 1 percent in these simulations). Since the change is assumed to persist, the price is maintained in these simulations as compared to the one time changes considered above. What is not apparent from this table is the substantial response which occurs on the supply side in almost all cases, though with substantial delay. Consequently, there is a tendency of supply and demand to readjust to a new equilibrium at a price which is only a little higher than the base solution price.

A persistent change in the rate of growth of GDP has substantially greater effects, as one would anticipate. The tendency for the price to rise to substantially higher levels that is particularly apparent for wool, cotton, and sugar reflects the failure of supply to make a full adjustment to the persistent increase in demand. This summary table does not illustrate the full range of dynamic responses since there are not only adjustments on the supply side but also reactions to the higher price on the side of demand. The full details of these reactions have been discussed in Chapter Five.

CONCLUDING REMARKS

The commodity models presented in this study are intended as a step in analyzing numerous commodity markets with econometric model structures. Commodity experts in each of the markets considered will be able to contribute information which may significantly modify each of the models. But we deem as highly promising the fact that simple models are able to describe many aspects of the behavior of the eight commodity markets over the sample period and that, by and large, these models have reasonable dynamic multiplier properties. The markets differ significantly, but it was possible to build models in each case and to observe dynamic behavior properties for these models which are substantially in accord with prior notions. While these models cannot be considered definitive in any sense, they reveal a surprising richness of response properties.

The models have important potentials for application even without further development, but there are significant possibilities for further consideration:

1. The role of various interventions in free markets. Significant interventions in commodity market occur in the producing and consuming countries and at the level of intergovernmental commodity

Table 6-6. Effect on Price of Exogenous 1 Percent Changes in GDP*

	Cocoa	Coffee	Tea	Wool	Cotton	Sugar	Wheat	Rice
1 Percent Persistent Increase in GDP								
Year 1	100.4	100.3	100.1	105.3	100.0	101.0	100.0	100.5
Year 2	100.6	100.2	100.2	103.6	100.2	101.8	100.2	101.2
Year 10	100.5	100.3	100.5	102.7	107.1	101.0	100.3	98.7
1 Percent Persistent Increase in Rate of Change of GDP								
Year 1	100.4	100.3	100.1	105.3	100.0	101.0	100.0	100.5
Year 2	101.0	100.4	100.4	109.5	100.3	103.0	100.3	101.7
Year 10	107.6	102.4	104.5	132.9	153.8	113.9	103.6	113.1

*multiplier simulation value/base simulation value × 100.0

organizations. These interventions include discrepancies between world market prices and prices paid by the consumer or received by the producer, production quotas, import restrictions, and various types of stockpiling and price stabilization schemes.

2. The role of expectations and speculation in the determination of commodity prices.
3. The role of nonlinearities in the structure of the models particularly with respect to price determination and inventories.
4. The impact of the weather.
5. The interaction between commodity prices and technological change on the supply and demand side.

Thus there remains substantial scope for further research on the models themselves.

However, the models offer considerable potential for application even without additional development. They provide the basis for relatively straightforward prediction, since they require only very limited projections for the exogenous variables. Second, they provide an initial framework for policy analysis. It is possible to introduce into the models modifications of supply and demand and inventories that approximate some potential policy measures and to get measurements of market responses under alternative policies. Finally, the models are well adapted for introduction into a framework of country models like Project LINK. Such a comprehensive interactive framework of models lends itself for the study of the world economy, its responses to inflation, the potentials for economic development, the impact of world business cycles, etc. The potentials in this direction are considerable and this volume is thus a step in the direction of comprehensive modeling of the world economy.

NOTES

1. Of course the argument for the inclusion of this variable is much stronger on the more disaggregated country level than on the level of aggregation of this study.

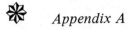 *Appendix A*

List of Variables

Appendix A.   List of Variables

Variable Name Common Variables	Variable Description	Sources and Notes
POP	Population, world, millions of people	"Production Year Book," FAO, 1972 for 1961–1972; pre-1961 figures extrapolated
POPA	Population, developed countries, millions of people	as above
POPB	Population, developing countries, millions of people	as above
POPC	Population, centrally planned countries, millions of people	as above
GDP	Gross domestic product index, world, services excluded, 1963 = 100	"Statistical Yearbook," United Nations, 1973, and previous issues, Table 3; GDP indexes are per capita
GDPA	Gross domestic product index, developed countries, services included, 1963 = 100	as above, Table 4
GDPB	Gross domestic product index, developing countries, services included, 1963 = 100	as above, Table 4
GDPC	Gross domestic product index, centrally planned countries, services excluded, 1963 = 100	as above, Table 3
DF	Price index of gross national product at market prices, OECD total, 1963 = 100 (used as U.N. export price index deflator)	"National Accounts of OECD Countries," 1973, and previous issues
T	Time Trend, 1947 = 1	
Specific Com-modity Variables: Cocoa		
PROW	Net world crop	"Cocoa Statistics," Gill & Duffus Group Ltd., London, 1973; obtained by adjusting the gross world crop for 1 percent loss in weight; thousand long tons
DW	Calendar year grindings, world	Grindings represent the absorption or disappearance of cocoa beans in each country—i.e., net imports of cocoa beans adjusted for changes in stocks; thousand long tons

DA	Calendar year grindings, developed countries, thousand long tons	As above
DB	Calendar year grindings, developing countries, thousand long tons	As above
DC	Calendar year grindings, centrally planned countries, thousand long tons	As above
STK	Closing world stocks, thousand long tons	Stocks for 1965 include an adjustment of 5000 tons of Nigerian cocoa beans sold for nontraditional uses and excluded from world grindings, but which are deducted from stocks
P	Export price index	United Nations
PDF	P deflated by DF	
S65	Supply dummy = 1 in 1965.	"Commodity Review & Outlook," FAO 33 percent increase in African production, about 10 percent of which due to favorable weather and remainder due to crop improvement factors such as selective breeding, pest and disease control, and the like
S7072	Supply dummy = 1 in 1970, 2 in 1971, 3 in 1972.	1970 was largest crop since 1965 record with larger crops in Africa, Brazil, and Dominican Republic; 1971, larger crops again, particularly in Brazil and Ecuador; 1972 still larger crops in Brazil and West Africa

Specific Commodity Variables: Coffee

PROW	Production, world, thousands of bags of 60 kilograms	"Annual Coffee Statistics," Pan American Coffee Bureau, 1972
DW	Demand, world	As above, calculated as: net exports (for consumption and working stocks in importing countries, where working stocks in importing countries are estimated to average four million bags) plus domestic distribution (for domestic consumption in producing countries); thousands of bags of 60 kilograms
DA	Demand, developed countries	As above, calculated as: percent of world imports of developed countries x world net exports plus domestic distribution (for grams
DB	Demand, developing countries	As above, calculated as: percent world imports of developing countries x world net exports plus domestic distribution (for domestic consumption in producing countries, all of which are developing); thousands of bags of 60 kilograms

continued

Appendix A continued

Common Variables Variable Name	Variable Description	Sources and Notes
DC	Demand, centrally planned countries, thousands of bags of 60 kilograms	As above, calculated as: percent world imports of centrally planned countries x world net exports
STK	Closing world stocks, thousands of bags of 60 kilograms	As above, but recalculated with 1947 base from above source to remove discrepancies caused by PACB revisions according to latest USDA figures (stocks are reduced in 1961 by three million bags which were allocated for industrial use in Brazil; Stocks are reduced in 1962 by seven million bags which were destroyed in Brazil in mid-1961)
P	Export price index, 1963 = 100	United Nations
PDF	P/DF	
S60	Supply dummy, 1 in 1960	"Commodity Review and Outlook," FAO Record harvest, largely reflecting a record Brazilian crop
S6566	Supply dummy, -1 in 1965, and 1 in 1966.	As above, world coffee harvest fell sharply in 1965 due to earlier drought, frosts, and fires in Brazil, the main producing country; record harvest in 1966 due to recovery of production in Brazil, combined with the upward trend in the rest of Latin America and in Africa
P6470	Price dummy, 3 in 1964, 2 in 1965, 1 in 1966, 2 in 1970.	As above, high, though successively declining prices in 1964, 1965, 1966 reflect a sharp fall in coffee harvest in 1964–1965 causing a pronounced rise in prices at the end of 1963 and continuing to 1964; Though somewhat lower than in 1964, a high level of prices was maintained in 1965 by use of export quotas, under International Coffee Agreement, with a new indicator price; coffee prices declined further in 1966 from levels to which they had recovered in 1964, but prices were still higher than in 1962 when the International Coffee Agreement was signed; in July of 1969, severe frost in Brazil followed by droughts caused a sharp upswing in price for 1970 with increased demands from importers for stocks

Specific Com-
modity Variables:
Tea

PROW — Production, world 1000 metric tons — "Production Year Book," 1972 and previous issues; and "Annual Bulletin of Statistics," International Tea Committee, 1973 and previous issues

PROA — Production, developed countries, 1000 metric tons — As above

PROB — Production, developing countries, 1000 metric tons — As above

PROC — Production, centrally planned countries, 1000 metric tons — As above

DW — Consumption, world, 1000 metric tons — As above; consumption figures are imports for consumption and consumption in producing countries

DA — Consumption, developed countries, 1000 metric tons — As above

DB — Consumption, developing countries, 1000 metric tons — As above

DC — Consumption, centrally planned countries, 1000 metric tons — As above

STK — Closing stocks, world, 1000 metric tons — Stock Base in 1954 estimated

P — Export price index, 1963 = 100 — United Nations

PDF — P/DF

Specific Com-
modity Variables:
Wool

PROW — Production, world, clean basis, 1000 metric tons — "Commodity Review & Outlook," FAO; and "UNCTAD Commodity Survey," 1966 (season beginning 1 October in Argentina and Uruguay; 1 July in other countries)

PROA — Production, developed countries, clean basis, 1000 metric tons — As above

*This is the source for all the descriptions of dummy variables below.

continued

Appendix A continued

Common Variables Variable Name	Variable Description	Sources and Notes
PROB	Production, developing countries, clean basis, 1000 metric tons	As above
PROC	Production, centrally planned countries, 1000 metric tons	As above
DW	Utilization, world, clean basis, 1000 metric tons	As above
DA	Utilization, developed countries, 1000 metric tons	As above
DB	Utilization, developing countries, 1000 metric tons	As above
DC	Utilization, centrally planned countries, 1000 metric tons	As above
STK	Closing stocks, world, 1000 metric tons	"Wool Intelligence," Commonwealth Economic Committee; stock series computed from 1972 beginning stock base as estimated world supply stocks of raw wool at beginning of season plus reported commercial stocks of raw wool in specified consuming countries at 1 January
P	Export price index, 1963 = 100	United Nations
PDF	P/DF	
PSYN	Index of manmade fibre textile products, 1967 = 100	"Survey of Current Business"
PW/PSYN	P/PSYN	Price during 1963–4
DUM 6364		

Specific Commodity Variables: Cotton

PROW	Production, world, 1000 metric tons	"Cotton—World Statistics," quarterly bulletin of International Cotton Advisory Committee (production figures are for the season 1 August–31 July)
PROA	Production, developed countries, 1000 metric tons	As above

PROB	Production, developing countries, 1000 metric tons	As above
PROC	Production, centrally planned countries, 1000 metric tons	As above
DW	Consumption, world, 1000 metric tons	As above (consumption figures are for the season 1 August–31 July)
DA	Consumption, developed countries, 1000 metric tons	As above
DB	Consumption, developing countries, 1000 metric tons	As above
DC	Consumption, centrally planned countries, 1000 metric tons	As above
STK	Closing stocks, world, 1000 metric tons	As above; stock series computed from a 31 July, 1973 base and include stocks afloat; stock adjustment of 300.0 in 1963 and 300.0 in 1964 reflects inadequacy of stock data
P	Export price index, 1963 = 100	United Nations
PDF	P/DF	
PSYN	Index of manmade fibre textile products, 1967 = 100	"Survey of Current Business"
PUSCT	Average spot price of United States cotton, U.S. cents/pound	"Commodity Year Book," Crop year beginning 1 August
GNPDF	United States GNP deflator, 1958 = 100	"Economic Report of the President"
PUSDF	PUSCT/GNPDF	
USALC	United States government acreage allotments for cotton, 1,000,000 acres	"Commodity Year Book"
PCSYN	P/PSYN	
DSB71	Production dummy, developing countries, 1971 = 1	"Commodity Review & Outlook," FAO; decline in production in most developing countries, notably in Latin America where poor profitability of cotton during 1969/1970 caused a switch to other crops

Specific Commodity Variables: Sugar

PROW	Production, world, 1,000,000 metric tons, raw value, sugar centrifugal raw	"Commodity Review & Outlook," FAO; *The World Sugar Economy, Structure & Policies*, International Sugar Council, 1963; *The World Sugar Economy in Figures, 1880–1959*, FAO, Commodity Reference Series
PROA	Production, developed countries, 1,000,000 metric tons	As above
PROB	Production, developing countries, 1,000,000 metric tons	As above

continued

Appendix A continued

Common Variables Variable Name	Variable Description	Sources and Notes
PROC	Production, centrally planned countries, 1,000,000 metric tons	As above
DW	Consumption, world, 1,000,000 metric tons, raw value, sugar centrifugal raw	As above; excluding consumption for nonfood use
DA	Consumption, developed countries, 1,000,000 metric tons	As above
DB	Consumption, developing countries, 1,000,000 metric tons	As above
DC	Consumption, centrally planned countries, 1,000,000 metric tons	As above
STK	Carryover stocks, 1,000,000 metric tons	"Commodity Review & Outlook," FAO; raw value, sugar centrifugal raw, world estimates, 31 August; stocks computed from 1969 base
P	Export price index, 1963 = 100	United Nations
PDF	P/DF	
TC	Time trend, 1955 = 1, increasing until 1964 and thereafter held at 10	Time trend specific to sugar production in the centrally planned countries
DUM 3	Production dummy, developing countries, $1961 = 1, 1963 = -1, 1970 = 1, 1972 = -1$	Dummy variable for crop variations in developing countries
DUMDS	Price dummy, $1956 = -1, 1963 = -1, 1972 = -1$	Dummy variable for high prices in 1957, 1964, and 1973
Specific Commodity Variables: Wheat		
PROW	Production, world, 1,000,000 metric tons	"Production Year Book," FAO; "Statistical Year Book," United Nations; (figures for 1957 and earlier are crop year basis; figures for 1958 and later are calendar year basis)
PROA	Production, developed countries, 1,000,000 metric tons	As above

PROB	Production, developing countries, 1,000,000 metric tons	As above
PROC	Production, centrally planned countries, 1,000,000 metric tons	As above
DW	Consumption, world, 1,000,000 metric tons	As above, computed as Production plus Net Imports−ΔSTK; export and imports of wheat and wheat flour are from "World Grain Trade Statistics," FAO (trade season: July–June); all data on flour are expressed in wheat equivalent; trade between EEC member countries is included; trade between centrally planned countries is not included as this trade is not reported on a July–June basis
DA	Consumption, developed countries, 1,000,000 metric tons	As above
DB	Consumption, developing countries, 1,000,000 metric tons	As above
DC	Consumption, centrally planned countries, 1,000,000 metric tons	As above
STK	End year carryover stocks; total seven exporting countries	"World Wheat Statistics," International Wheat Council, 1973; stocks computed from 1971–1972 season base
P	Export price index, 1963 = 100	United Nations
PDF	P/DF	
IMPC	Imports, centrally planned countries, 1,000,000 metric tons	Source and note for trade figures given above.
SA610	Production dummy, developed countries, 1961 = 1, 1970 = 1	"Commodity Review & Outlook, FAO; in 1961, combined output of United States and Canada fell by 20 percent as a result of prolonged drought during growing season; in 1970, lower production in developed countries mainly reflected government policy to reduce output and carryover stocks in Australia, Canada, & U.S.
SB66	Production dummy, developing countries, 1966 = 1	As above; in 1966, production dropped by more than half in North Africa, India, Pakistan, and most of Near East due to severe drought
SC63	Production dummy, centrally planned countries, 1963 = 1	As above; in 1963, heavy crop losses occurred in USSR due to the combination of a long and exceptionally cold winter and a late and excessively dry summer

continued

Appendix A continued

Common Variables Variable Name	Variable Description	Sources and Notes
SC66	Production dummy, centrally planned countries, 1966 = 1	As above; in 1966, USSR production increased to 100 million metric tons as a result of the combined effect of improved farming techniques, larger inputs of fertilizer and other resources, and very favorable weather
Specific Commodity Variables: Rice		
PROW	Production, world, milled equivalent converted from paddy at 65 percent, 1,000,000 metric tons	"Production Year Book," FAO, 1972 for 1961–1972 and "The World Rice Economy in Figures, 1909–1963," FAO, for pre-1961 figures; the former are on a calendar year basis, the latter on a crop year basis
PROA	Production, developed countries, milled equivalent converted from paddy at 65 percent, 1,000,000 metric tons	As above
PROB	Production, developing countries, milled equivalent converted from paddy at 65 percent, 1,000,000 metric tons	As above
PROC	Production, centrally planned countries, milled equivalent converted from paddy at 65 percent, 1,000,000 metric tons	As above
DW	Consumption, world, 1,000,000 metric tons, milled equivalent	As above; and "Commodity Review & Outlook," FAO for exports, indigenous, and imports retained, 1962–1972; pre-1962 import and export figures refer to general trade and are from the "World Rice Economy in Figures, 1909–1963," FAO; consumption figures are Production plus Net Imports— ΔStk
DA	Consumption, developed countries, 1,000,000 metric tons, milled equivalent	As above

DB	Consumption, developing countries, 1,000,000 metric tons, milled equivalent	As above
DC	Consumption, centrally planned countries, 1,000,000 metric tons, milled equivalent	As above
STK	Carryover stocks, selected exporting and importing countries, augmented by 10 percent of said 1971 stock level in each year, 1,000,000 metric tons, milled equivalent	"FAO Rice Report," FAO, 1972 and earlier
P	Export price index, 1963 = 100	United Nations
PDF	P/DF	
SB756	Production dummy, developing countries, 1975 = 1, 1965 = 1, 1966 = 1	"Commodity Review & Outlook," FAO; the rice crop of South and Southeast Asia harvested late in 1957 and early in 1958 suffered from lack of rain over large areas. Adverse conditions, either flood or drought, cut production in 1956 and 1966 in certain parts of Asia
DB656	Consumption dummy, developing countries, 1965 = 1, 1966 = 1	As above; Asian shortages of rice could not be met by rice imports alone and were largely provided by massive imports of wheat.
DC59	Consumption dummy, centrally planned countries, 1959 = 1	As above; in 1959–1960 prolonged drought and floods considerably reduced mainland Chinese rice crop; this, coupled with record rice exports of 1.6 million tons in 1959, left a reduced amount of rice available for home consumption
P678	Price dummy, 1967 = 1, 1968 = 1	As above; continuing heavy pressure of demand on limited supplies caused a further sharp increase in rice prices in 1967 which did not ease off until 1970
P70	Price dummy, 1970 = 1	As above; export supplies were considerably higher in 1970 and import demand was limited owing to better crops in many importing countries; consequently, rice prices declined steeply

Appendix B

Data (definitions and sources are given in Appendix A)

VARIABLES COMMON TO ALL COMMODITIES

Year	POP	POPA	POPB	POPC	GDP	GDPA	GDPB	GDPC	DF	T
1947-	0.0	0.0	0.0	0.0	0.0	0.0	0.0	0.0	62.000	1.000
1948-	0.0	0.0	0.0	0.0	0.0	0.0	0.0	0.0	67.000	2.000
1949-	0.0	0.0	0.0	0.0	0.0	0.0	0.0	0.0	67.000	3.000
1950-	0.0	0.0	0.0	0.0	0.0	0.0	0.0	0.0	67.000	4.000
1951-	0.0	0.0	0.0	0.0	0.0	0.0	0.0	0.0	73.000	5.000
1952-	2591.408	586.288	1129.545	892.874	71.000	75.000	78.000	50.000	76.000	6.000
1953-	2591.408	599.766	1144.026	892.874	74.000	76.000	81.000	53.000	77.000	7.000
1954-	2686.216	599.766	1172.989	903.257	75.000	78.000	84.000	57.000	78.000	8.000
1955-	2717.818	606.505	1187.470	913.639	80.000	82.000	85.000	63.000	80.000	9.000
1956-	2781.024	613.244	1230.914	924.021	82.000	84.000	87.000	67.000	83.000	10.000
1957-	2812.626	626.721	1259.877	944.786	84.000	85.000	88.000	72.000	86.000	11.000
1958-	2875.831	633.460	1288.839	955.168	85.000	84.000	90.000	79.000	89.000	12.000
1959-	2939.036	640.199	1303.321	975.932	88.000	88.000	92.000	82.000	91.000	13.000
1960-	2970.639	653.677	1317.802	996.697	93.000	90.000	95.000	89.000	93.000	14.000
1961-	3039.900	657.532	1378.575	1003.794	95.000	95.000	96.000	94.000	95.000	15.000
1962-	3099.586	665.795	1412.908	1020.883	98.000	97.000	98.000	97.000	97.400	16.000
1963-	3160.254	673.894	1448.134	1038.226	100.000	100.000	100.000	100.000	100.000	17.000
1964-	3221.841	681.822	1484.617	1055.402	105.000	105.000	104.000	107.000	102.800	18.000
1965-	3284.583	689.559	1522.533	1072.490	109.000	109.000	107.000	114.000	105.900	19.000
1966-	3348.127	696.813	1561.905	1089.409	114.000	115.000	110.000	121.000	109.500	20.000
1967-	3412.691	703.593	1602.672	1106.426	116.000	118.000	113.000	130.000	113.000	21.000
1968-	3479.125	710.291	1644.947	1123.887	121.000	123.000	118.000	138.000	116.900	22.000
1969-	3547.385	717.456	1688.813	1141.115	126.000	128.000	123.000	144.000	122.600	23.000
1970-	3617.362	724.913	1734.169	1158.279	129.000	130.000	126.000	153.000	129.800	24.000
1971-	3688.176	731.563	1781.010	1175.603	132.000	134.000	130.000	162.000	134.600	25.000
1972-	3760.745	738.518	1829.240	1192.986	136.000	139.000	130.000	169.000	144.300	26.000
1973-	3834.135	744.685	1878.823	1210.625	142.000	148.000	133.000	181.000	160.600	27.000

COMMODITY SPECIFIC VARIABLES: COCOA

	PROW	DW	DA	DB	DC	STK	P	PDF
1947-	608.000	650.000	0.0	0.0	0.0	158.000	106.000	1.710
1948-	587.000	617.000	0.0	0.0	0.0	128.000	149.000	2.224
1949-	765.000	678.000	0.0	0.0	0.0	215.000	85.000	1.269
1950-	748.000	760.000	0.0	0.0	0.0	203.000	106.000	1.582
1951-	795.000	757.000	630.000	106.000	21.000	241.000	139.000	1.904
1952-	636.000	725.000	605.000	97.000	23.000	152.000	138.000	1.816
1953-	790.000	777.000	622.000	129.000	26.000	165.000	132.000	1.714
1954-	768.000	752.000	609.000	111.000	32.000	181.000	219.000	2.808
1955-	794.000	725.000	571.000	116.000	38.000	250.000	142.000	1.775
1956-	835.000	798.000	632.000	127.000	39.000	287.000	103.000	1.241
1957-	889.000	887.000	701.000	137.000	49.000	289.000	119.000	1.384
1958-	765.000	865.000	650.000	157.000	58.000	189.000	171.000	1.921
1959-	899.000	860.000	626.000	169.000	65.000	228.000	138.000	1.516
1960-	1029.000	907.000	672.000	162.000	73.000	350.000	107.000	1.151
1961-	1161.000	988.000	736.000	169.000	83.000	523.000	86.000	0.905
1962-	1113.000	1083.000	806.000	183.000	94.000	553.000	81.000	0.832
1963-	1152.000	1133.000	842.000	177.000	114.000	572.000	100.000	1.000
1964-	1204.000	1170.000	846.000	191.000	133.000	606.000	91.000	0.885
1965-	1467.000	1280.000	905.000	227.000	148.000	788.000	66.000	0.623
1966-	1193.000	1354.000	931.000	252.000	171.000	627.000	94.000	0.858
1967-	1320.000	1366.000	915.000	272.000	179.000	581.000	110.000	0.973
1968-	1320.000	1384.000	898.000	298.000	188.000	517.000	133.000	1.138
1969-	1209.000	1347.000	865.000	299.000	183.000	379.000	174.000	1.419
1970-	1404.000	1335.000	842.000	304.000	189.000	448.000	127.000	0.978
1971-	1464.000	1398.000	862.000	327.000	209.000	514.000	100.000	0.743
1972-	1546.000	1496.000	896.000	361.000	239.000	564.000	117.000	0.811
1973-	-50.000	1589.000	969.000	361.000	259.000	337.000	220.000	1.370

COMMODITY SPECIFIC VARIABLES: COFFEE

	PROW	DW	DA	DB	DC	STK	P.	PDF
1947-	35308.000	34648.000	0.0	0.0	0.0	17050.000	74.000	1.194
1948-	34618.000	39140.000	0.0	0.0	0.0	12528.000	77.000	1.149
1949-	39095.000	41596.000	0.0	0.0	0.0	10027.000	91.000	1.358
1950-	37615.000	39599.000	0.0	0.0	0.0	3043.000	138.000	2.060
1951-	38164.000	39756.000	0.0	0.0	0.0	6451.000	154.000	2.110
1952-	38530.000	39798.000	0.0	0.0	0.0	5183.000	155.000	2.039
1953-	41513.000	41175.000	0.0	0.0	0.0	5521.000	164.000	2.130
1954-	43996.000	43114.000	31618.000	11262.000	234.000	6403.000	214.000	2.744
1955-	42188.000	37485.000	27641.000	9639.000	205.000	11106.000	156.000	1.950
1956-	50348.000	45027.000	36151.000	8531.000	345.000	16427.000	165.000	1.988
1957-	45420.000	46981.000	34176.000	12552.000	253.000	14866.000	158.000	1.837
1958-	55009.000	46119.000	34987.000	10609.000	523.000	23756.000	139.000	1.562
1959-	61665.000	48641.000	36755.000	11184.000	702.000	36780.000	110.000	1.209
1960-	78919.000	54849.000	39641.000	14319.000	889.000	60850.000	105.000	1.129
1961-	65768.000	57174.000	41213.000	14988.000	973.000	66444.000	101.000	1.063
1962-	72043.000	59129.000	42458.000	15537.000	1134.000	72358.000	97.000	0.996
1963-	67404.000	63871.000	46669.000	15906.000	1296.000	75891.000	100.000	1.000
1964-	70098.000	63467.000	45470.000	16417.000	1580.000	83422.000	133.000	1.294
1965-	50613.000	58181.000	39883.000	16647.000	1651.000	75854.000	124.000	1.171
1966-	81604.000	67002.000	47098.000	17942.000	1962.000	90456.000	120.000	1.096
1967-	60577.000	66621.000	46297.000	18818.000	2070.000	84412.000	112.000	0.991
1968-	68612.000	70559.000	49487.000	19786.000	2254.000	82465.000	112.000	0.958
1969-	61068.000	70851.000	48352.000	20439.000	2713.000	72682.000	117.000	0.954
1970-	66635.000	72301.000	49062.000	20240.000	2800.000	67016.000	153.000	1.179
1971-	58300.000	70500.000	47802.000	21358.000	2458.000	54816.000	134.000	0.996
1972-	71800.000	76500.000	52152.000	22629.000	2990.000	50116.000	150.000	1.040
1973-	5700.000	79354.000	53465.000		3260.000	48035.000	188.430	1.173

COMMODITY SPECIFIC VARIABLES: TEA

Year	PROW	PRDA	PROB	PROC	DW	DA	DB	DC	STK	Pi	PDF
1947-	0.0	0.0	0.0	0.0	0.0	0.0	0.0	0.0	0.0		1.274
1948-	0.0	0.0	0.0	0.0	0.0	0.0	0.0	0.0	0.0	79.000	1.194
1949-	0.0	0.0	0.0	0.0	0.0	0.0	0.0	0.0	0.0	80.000	1.433
1950-	0.0	0.0	0.0	0.0	0.0	0.0	0.0	0.0	0.0	96.000	1.567
1951-	0.0	0.0	0.0	0.0	0.0	0.0	0.0	0.0	0.0	105.000	1.342
1952-	628.000	57.000	495.000	76.000	0.0	0.0	0.0	44.400	0.0	98.000	1.039
1953-	614.000	57.000	465.000	92.000	669.100	394.700	219.000	55.400	2726.800	79.000	1.169
1954-	780.000	68.000	598.000	114.000	692.000	418.400	212.200	61.400	2814.800	90.000	1.705
1955-	830.000	73.000	627.000	130.000	755.300	404.400	262.700	88.200	2889.500	133.000	1.587
1956-	849.000	71.000	628.000	150.000	759.700	423.300	235.800	100.600	2978.800	127.000	1.361
1957-	950.000	72.000	736.000	142.000	814.900	430.800	282.100	102.000	3113.900	113.000	1.221
1958-	910.000	75.000	657.000	178.000	860.200	437.600	295.600	127.000	3163.700	105.000	1.180
1959-	940.000	79.000	670.000	191.000	876.000	436.100	297.900	142.000	3227.700	105.000	1.143
1960-	970.000	78.000	691.000	201.000	897.800	440.900	311.800	145.100	3299.900	104.000	1.129
1961-	1045.000	81.000	763.000	201.000	934.800	455.400	334.900	149.500	3410.100	105.000	1.063
1962-	1062.000	77.000	779.000	206.000	943.600	452.000	342.500	149.100	3528.500	101.000	1.078
1963-	1082.000	81.000	791.000	210.000	956.200	461.500	341.600	153.100	3654.300	105.000	1.000
1964-	1127.000	83.000	834.000	210.000	1002.300	473.600	364.300	164.400	3779.000	100.000	0.982
1965-	1145.000	77.000	856.000	212.000	1001.900	457.000	378.900	166.000	3922.100	101.000	0.954
1966-	1186.000	83.000	881.000	224.000	1041.303	461.700	427.100	152.500	4066.800	101.000	0.950
1967-	1186.000	85.000	877.000	224.000	1060.100	480.600	418.700	160.800	4192.699	104.000	0.920
1968-	1235.000	85.000	924.000	226.000	1071.033	487.200	429.800	154.700	4356.000	91.000	0.778
1969-	1251.000	90.000	929.000	232.000	1123.000	475.300	479.000	168.700	4484.000	88.000	0.718
1970-	1297.000	91.000	936.000	270.000	1171.300	485.600	483.700	202.000	4609.699	84.000	0.647
1971-	1401.000	93.000	977.000	331.000	1250.500	496.900	485.400	268.200	4760.199	83.000	0.617
1972-	1526.000	95.000	1066.000	366.000	1321.000	490.700	522.100	308.200	4965.199	81.000	0.561
1973-	1570.000	95.000	1090.000	385.000	1341.000	502.500	517.600	321.000	5194.102	83.000	0.517

COMMODITY SPECIFIC VARIABLES: WOOL

	PROW	PROA	PROB	PROC	DW	DA	DB	DC	STK	P	PDF
1947-	0.0	0.0	0.0	0.0	0.0	0.0	0.0	0.0	0.0	0.0	0.0
1948-	0.0	0.0	0.0	0.0	0.0	0.0	0.0	0.0	0.0	0.0	0.0
1949-	0.0	0.0	0.0	0.0	0.0	0.0	0.0	0.0	0.0	0.0	0.0
1950-	1057.000	0.0	0.0	0.0	0.0	0.0	0.0	0.0	0.0	134.000	2.000
1951-	1069.000	0.0	0.0	0.0	0.0	0.0	0.0	0.0	0.0	173.000	2.370
1952-	1158.000	711.000	357.000	90.000	1069.000	816.000	109.000	144.000	182.000	101.000	1.329
1953-	1170.000	579.000	320.000	271.000	1114.000	821.000	125.000	168.000	283.000	114.000	1.481
1954-	1194.000	728.000	274.000	192.000	1209.000	857.000	122.000	230.000	363.000	108.000	1.385
1955-	1263.000	770.000	279.000	214.000	1300.000	927.000	119.000	254.000	417.000	99.000	1.237
1956-	1338.000	834.000	281.000	223.000	1339.000	938.000	122.000	279.000	455.000	102.000	1.229
1957-	1310.000	785.000	285.000	243.000	1253.000	842.000	112.000	299.000	426.000	113.000	1.314
1958-	1383.000	840.000	282.000	261.000	1426.000	975.000	126.000	325.000	556.000	79.000	0.888
1959-	1461.000	892.000	280.000	289.000	1502.000	1001.000	138.000	356.000	591.000	84.000	0.923
1960-	1463.000	884.000	286.000	293.000	1511.000	1010.000	148.000	359.000	552.000	85.000	0.914
1961-	1488.000	906.000	292.000	290.000	1511.000	1014.000	141.000	349.000	526.000	84.000	0.884
1962-	1481.000	922.000	268.000	291.000	1513.000	1019.000	142.000	381.000	496.000	34.000	0.862
1963-	1513.000	949.000	270.000	294.000	1513.000	976.000	156.000	404.000	499.000	100.000	1.000
1964-	1537.000	933.000	292.000	312.000	1554.000	995.000	155.000	411.000	523.000	103.000	1.002
1965-	1529.000	921.000	295.000	313.000	1620.000	1037.000	172.000	428.000	498.000	86.000	0.812
1966-	1560.000	942.000	295.000	323.000	1560.000	916.000	215.000	462.000	438.000	90.000	0.822
1967-	1594.000	913.000	339.000	342.000	1655.000	969.000	224.000	446.000	472.000	77.000	0.681
1968-	1680.000	967.000	354.000	359.000	1698.000	1009.000	243.000	465.000	497.000	74.000	0.633
1969-	1669.000	975.000	351.000	345.000	1665.000	944.000	256.000	516.000	468.000	73.000	0.595
1970-	1654.000	943.000	351.000	361.000	1640.000	905.000	219.000	496.000	457.000	63.000	0.485
1971-	1593.100	923.500	310.200	354.400	1701.000	986.000	219.000	516.000	410.000	57.000	0.423
1972-	1568.199	909.100	308.500	350.600	1635.000	0.0	0.0	496.000	277.300	88.000	0.610
1973-	1470.500	808.100	305.000	357.400	0.0	0.0	0.0	0.0	112.800	183.000	1.139

COMMODITY SPECIFIC VARIABLES: COTTON

	PROW	PROA	PROB	PROC	DW	DA	DB	DC	STK	PC
1947-	4749.000	1893.000	1891.000	965.000	6109.000	3594.000	1343.000	1172.000	3871.000	151.000
1948-	5518.000	2589.000	1889.000	1040.000	6461.000	3577.000	1383.000	1501.000	2928.000	181.000
1949-	6474.000	3251.000	2016.000	1207.000	6316.000	3358.000	1428.000	1530.000	3086.000	179.000
1950-	7148.000	3521.000	2313.000	1314.000	6720.000	3766.000	1369.000	1585.000	3514.000	162.000
1951-	6647.000	2210.000	2660.000	1777.000	7612.000	4357.000	1393.000	1862.000	2549.000	225.000
1952-	8390.000	3333.000	2888.000	2169.000	7625.000	3955.000	1518.000	2152.000	3314.000	169.000
1953-	8693.000	3334.000	2948.000	2411.000	8012.000	4019.000	1596.000	2397.000	3995.000	133.000
1954-	9061.000	3646.000	2951.000	2464.000	8437.000	4091.000	1756.000	2590.000	4619.000	144.000
1955-	8917.000	3060.000	3376.000	2481.000	8655.000	4073.000	1915.000	2667.000	4881.000	138.000
1956-	9492.000	3327.000	3435.000	2730.000	8960.000	4161.000	2038.000	2761.000	5413.000	130.000
1957-	9221.000	3021.000	3374.000	2826.000	9339.000	4283.000	2147.000	2909.000	5295.000	121.000
1958-	9030.000	2507.000	3569.000	2954.000	9323.000	4021.000	2158.000	3144.000	5002.000	106.000
1959-	9738.000	2635.000	3697.000	3406.000	9924.000	4049.000	2277.000	3598.000	4816.000	94.000
1960-	10267.000	3319.000	3472.000	3476.000	10499.000	4400.000	2377.000	3722.000	4584.000	101.000
1961-	10113.000	3268.000	3962.000	2883.000	10206.000	4382.000	2487.000	3337.000	4491.000	101.000
1962-	9819.000	3352.000	4003.000	2464.000	9976.000	4435.000	2642.000	2899.000	4334.000	100.000
1963-	10457.000	3471.000	4536.000	2450.000	9802.000	4234.000	2657.000	2911.000	4989.000	100.000
1964-	10957.000	3566.000	4587.000	2804.000	10345.000	4342.000	2822.000	3181.000	5601.000	100.000
1965-	11502.000	3500.000	4788.000	3214.000	11318.000	4490.000	2974.000	3554.000	6085.000	97.000
1966-	11884.000	3479.000	4906.000	3499.000	11278.000	4535.000	2964.000	3779.000	6691.000	94.000
1967-	10818.000	2325.000	4733.000	3760.000	11604.000	4502.000	3087.000	4015.000	5905.000	100.000
1968-	10689.000	1865.000	4961.000	3863.000	11666.000	4311.000	3256.000	4099.000	4928.000	102.000
1969-	11816.000	2633.000	5494.000	3689.000	11585.000	4204.000	3384.000	4097.000	5059.000	98.000
1970-	11366.000	2446.000	5375.000	3545.000	11781.000	4120.000	3549.000	4112.000	4644.000	103.000
1971-	11685.000	2460.000	4864.000	4361.000	12202.000	4098.000	3603.000	4496.000	4127.000	113.000
1972-	12445.000	2543.000	5813.000	4489.000	12541.000	4111.000	3775.000	4655.000	4431.000	132.000
1973-	13364.000	3272.000	5816.000	4276.000	12869.000	4073.000	4062.000	4734.000	4926.000	200.800

	PDF	PSYN	PUSCT	GNPDF	PUSDF	USALC	PCSYN
1947-	2.435	137.700	35.070	74.640	0.470	0.0	1.097
1948-	2.701	154.500	35.440	79.570	0.445	0.0	1.172
1949-	2.672	135.700	32.710	79.120	0.413	0.0	1.319
1950-	2.418	135.800	32.650	80.160	0.407	0.0	1.193
1951-	3.082	138.300	43.230	85.640	0.505	21.555	1.627
1952-	2.224	126.700	39.940	87.450	0.457	0.0	1.334
1953-	1.727	124.200	35.320	88.330	0.400	0.0	1.071
1954-	1.846	122.200	34.360	89.630	0.383	21.420	1.178
1955-	1.725	123.500	35.020	90.860	0.385	18.159	1.117
1956-	1.566	116.100	35.450	93.990	0.377	17.437	1.120
1957-	1.407	116.900	33.530	97.490	0.344	17.675	1.035
1958-	1.191	114.500	34.390	100.000	0.344	17.638	0.926
1959-	1.033	115.600	34.470	101.660	0.339	17.399	0.813
1960-	1.086	112.700	31.930	103.290	0.309	17.593	0.896
1961-	1.063	108.000	30.960	104.620	0.296	18.522	0.935
1962-	1.027	108.600	33.670	105.780	0.318	18.202	0.921
1963-	1.000	108.600	33.520	107.170	0.313	16.400	0.921
1964-	0.973	110.800	33.180	108.850	0.305	16.313	0.903
1965-	0.916	109.800	30.730	110.860	0.277	16.278	0.883
1966-	0.858	103.500	29.600	113.940	0.263	16.281	0.908
1967-	0.885	100.000	22.080	117.590	0.188	16.271	1.000
1968-	0.873	105.000	24.830	122.300	0.203	16.280	0.971
1969-	0.799	106.600	22.900	128.200	0.179	16.271	0.919
1970-	0.794	102.000	22.150	135.240	0.164	16.280	1.010
1971-	0.840	100.800	23.550	141.600	0.166	17.228	1.121
1972-	0.915	108.000	31.520	146.100	0.216	11.618	1.222
1973-	1.250	121.800	33.140	154.300	0.215	11.618	1.649

COMMODITY SPECIFIC VARIABLES: SUGAR

	PROW	PROA	PROB	PROC	DW	DA	DB	DC	STK	P	PDF
1947-	0.0	0.0	0.0	0.0	0.0	0.0	0.0	0.0	0.0	0.0	0.0
1948-	0.0	0.0	0.0	0.0	0.0	0.0	0.0	0.0	0.0	0.0	0.0
1949-	0.0	0.0	0.0	0.0	0.0	0.0	0.0	0.0	0.0	0.0	0.0
1950-	29.200	8.799	15.491	4.910	29.404	17.376	7.619	4.409	5.986	79.000	117.910
1951-	33.600	10.907	16.553	6.140	31.547	17.775	8.591	5.181	8.039	90.000	123.288
1952-	36.100	9.979	19.381	6.740	33.436	18.154	9.228	6.054	10.703	76.000	100.000
1953-	35.000	10.177	18.033	6.790	35.770	19.179	10.323	6.268	9.933	65.000	84.416
1954-	38.800	12.521	18.419	7.860	37.294	19.609	11.023	6.662	11.439	63.000	80.769
1955-	38.400	12.393	19.357	6.650	38.773	20.251	11.637	6.885	11.066	62.000	77.500
1956-	39.700	12.398	19.872	7.430	41.507	21.268	12.457	7.782	9.259	65.000	78.313
1957-	41.600	12.099	21.591	7.910	42.615	21.661	12.755	8.244	8.244	82.000	95.349
1958-	44.400	12.864	22.606	8.930	44.824	22.265	13.709	8.850	7.820	63.000	70.787
1959-	50.100	14.594	24.406	11.100	46.800	23.000	13.630	10.200	11.120	58.000	63.736
1960-	49.800	13.694	25.646	10.460	49.700	23.600	14.000	12.100	11.220	58.000	62.366
1961-	54.900	16.610	26.870	11.420	52.700	23.980	15.220	13.500	13.420	55.000	57.895
1962-	51.800	14.810	24.750	12.240	53.000	24.090	15.410	13.500	12.220	58.000	59.548
1963-	50.540	15.270	23.550	11.720	53.900	24.320	16.550	13.030	8.860	100.000	100.000
1964-	55.250	17.800	25.350	12.100	55.490	25.590	16.340	13.560	8.620	79.000	76.848
1965-	66.080	19.810	29.780	17.160	59.690	26.430	17.810	15.450	15.010	53.000	50.047
1966-	62.790	17.760	29.710	15.250	61.660	26.910	18.650	16.100	16.140	53.000	48.402
1967-	64.810	18.740	30.310	15.740	63.020	26.830	19.910	16.280	17.930	55.000	48.673
1968-	66.480	19.740	29.350	17.390	65.390	27.930	20.440	16.970	19.020	56.000	47.904
1969-	68.470	20.570	30.990	16.910	68.190	28.360	21.970	17.860	19.300	66.000	53.834
1970-	72.640	20.810	35.580	16.250	70.550	29.490	23.240	17.820	21.390	69.000	53.159
1971-	72.130	20.630	34.560	16.920	73.230	29.980	24.920	18.330	20.290	75.000	55.721
1972-	72.150	23.230	32.910	16.010	75.030	30.460	25.590	18.980	17.410	97.000	67.221
1973-	76.540	22.980	36.760	16.800	76.890	30.840	26.440	19.610	17.060	127.000	79.078

COMMODITY SPECIFIC VARIABLES: WHEAT

	PROW	PROA	PROB	PKOC	DW	DA	DB	DC	STK	P'	PDF	IMPC
1947-	0.0	0.0	0.0	0.0	0.0	0.0	0.0	0.0	0.0	0.0	0.0	0.0
1948-	0.0	0.0	0.0	0.0	0.0	0.0	0.0	0.0	0.0	0.0	0.0	0.0
1949-	0.0	0.0	0.0	0.0	0.0	0.0	0.0	0.0	0.0	0.0	0.0	0.0
1950-	0.0	0.0	0.0	0.0	0.0	0.0	0.0	0.0	0.0	0.0	1.612	0.0
1951-	0.0	0.0	0.0	0.0	0.0	0.0	0.0	0.0	0.0	108.000	1.589	0.0
1952-	205.000	94.800	36.200	74.000	188.400	73.400	42.100	72.900	16.900	116.000	1.526	0.0
1953-	203.600	91.800	40.900	70.900	185.900	71.700	43.900	70.300	34.600	116.000	1.558	0.0
1954-	194.800	78.300	40.900	75.600	191.000	73.500	41.700	75.800	38.400	120.000	1.346	0.300
1955-	206.500	85.100	39.000	82.400	204.200	77.200	43.400	83.600	40.700	105.000	1.225	1.100
1956-	226.500	81.000	42.400	103.100	224.400	72.400	49.500	102.500	42.800	98.000	1.157	1.700
1957-	221.500	82.100	44.200	95.200	225.200	76.700	53.100	95.400	39.100	96.000	1.116	0.600
1958-	252.700	97.500	42.600	112.600	241.700	78.200	51.900	111.600	50.100	96.000	1.079	1.100
1959-	243.700	93.000	43.200	107.500	243.100	79.300	56.300	107.400	50.700	96.000	1.044	0.600
1960-	243.600	100.600	43.300	99.700	240.000	80.200	57.400	102.400	54.300	95.000	1.011	1.000
1961-	228.500	88.600	43.100	96.800	236.500	76.500	58.900	101.000	46.300	94.000	1.021	4.000
1962-	259.000	104.600	48.700	106.300	257.500	82.000	64.400	111.100	47.800	97.000	1.037	5.600
1963-	239.600	103.300	51.000	85.300	246.700	78.900	64.500	103.300	40.700	101.000	1.000	6.800
1964-	277.100	110.900	51.400	114.800	275.300	83.200	66.600	123.200	42.500	100.000	1.012	18.700
1965-	267.400	111.800	50.900	104.700	278.100	86.400	68.500	125.500	31.800	104.000	0.897	11.300
1966-	310.100	117.300	48.200	144.600	307.900	84.000	70.100	153.700	34.000	95.000	0.932	18.600
1967-	298.900	119.500	52.900	126.500	290.100	83.300	76.700	130.100	42.800	102.000	0.912	10.500
1968-	331.200	129.600	59.900	141.700	312.900	88.900	77.500	144.500	63.100	103.000	0.830	6.700
1969-	315.200	121.300	64.600	129.300	312.100	93.100	83.600	135.400	66.200	97.000	0.767	5.100
1970-	318.400	103.700	64.400	150.300	333.000	91.600	87.200	154.900	51.600	94.000	0.693	7.600
1971-	353.900	126.300	71.200	150.300	354.400	96.300	95.300	162.800	51.100	90.000	0.713	5.000
1972-	347.000	122.000	78.400	146.600	366.600	96.000	101.430	169.200	31.500	110.000	0.762	22.800
1973-	377.500	133.200	71.200	173.100	385.300	100.800	102.600	181.900	23.700	214.000	1.333	11.400

COMMODITY SPECIFIC VARIABLES: RICE

	PRUW	PROA	PROB	PROC.	DW	DA	DB	DC.	STK.	P	PDF
1947-	0.0	0.0	0.0	0.0	0.0	0.0	0.0	0.0	0.0	0.0	0.0
1948-	0.0	0.0	0.0	0.0	0.0	0.0	0.0	0.0	0.0	0.0	0.0
1949-	0.0	0.0	0.0	0.0	0.0	0.0	0.0	0.0	0.0	0.0	0.0
1950-	0.0	0.0	0.0	0.0	0.0	0.0	0.0	0.0	0.0	0.0	1.478
1951-	107.600	0.0	0.0	0.0	0.0	0.0	0.0	0.0	0.0	99.000	1.521
1952-	116.000	10.800	0.0	0.0	0.0	0.0	0.0	0.0	0.0	111.000	1.605
1953-	126.000	9.600	71.900	44.600	0.0	0.0	0.0	0.0	0.0	122.000	1.727
1954-	123.400	10.600	68.300	44.500	0.0	0.0	0.0	44.300	0.0	133.000	1.487
1955-	133.400	13.100	70.700	49.500	0.0	0.0	0.0	49.600	0.100	116.000	1.337
1956-	140.500	11.800	76.100	52.700	138.700	0.0	75.900	52.500	1.900	107.000	1.193
1957-	137.300	11.800	70.400	55.100	138.300	10.500	70.700	55.000	20.990	99.000	1.163
1958-	147.200	12.500	78.500	56.200	147.100	12.500	78.900	55.300	21.090	100.000	1.146
1959-	147.200	13.100	82.000	52.100	146.500	12.800	82.100	51.400	21.790	102.000	1.022
1960-	155.300	13.400	87.400	54.500	154.900	13.000	87.500	54.200	22.190	93.000	0.957
1961-	157.700	13.100	87.900	56.600	157.900	13.100	88.600	56.500	21.990	89.000	0.979
1962-	159.400	14.000	87.900	57.500	159.900	12.700	88.600	57.500	21.490	93.000	1.016
1963-	165.900	13.800	94.400	57.700	165.700	13.800	94.700	57.500	21.690	99.000	1.000
1964-	172.600	13.800	98.500	60.300	172.600	13.400	98.800	60.100	21.690	100.000	0.973
1965-	166.800	13.700	90.000	63.200	166.400	13.700	90.000	63.000	22.090	103.000	0.973
1966-	165.600	14.300	88.900	62.500	165.300	13.500	89.800	61.800	22.390	108.000	0.986
1967-	180.400	15.900	99.300	65.100	178.600	13.800	100.600	64.500	24.190	125.000	1.106
1968-	185.100	16.300	104.200	64.500	183.300	13.700	105.500	64.300	25.990	137.000	1.172
1969-	190.800	15.800	107.400	67.500	187.600	11.700	108.400	67.300	29.190	136.000	1.109
1970-	200.700	14.400	114.100	72.200	198.100	11.000	115.500	71.700	31.790	100.000	0.770
1971-	199.100	12.900	113.000	73.200	199.600	11.900	114.900	72.800	31.290	97.000	0.721
1972-	190.300	13.600	106.300	70.400	195.000	14.600	113.400	70.300	26.590	108.000	0.748
1973-	208.400	14.300	119.100	75.000	211.400	14.200	123.700	73.500	23.590	198.000	1.233

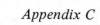 *Appendix C*

Charts of Dynamic Simulations

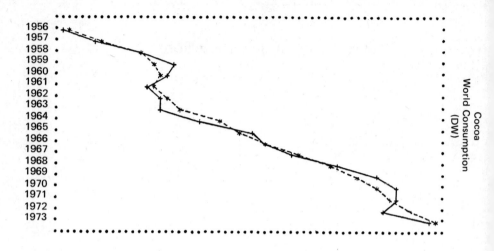

Cocoa
World Consumption
(DW)

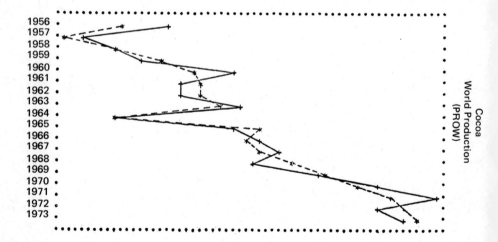

Cocoa
World Production
(PROW)

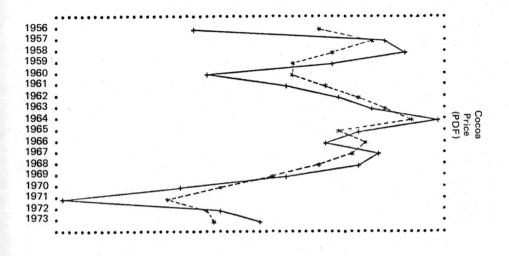

Cocoa
Price
(PDF)

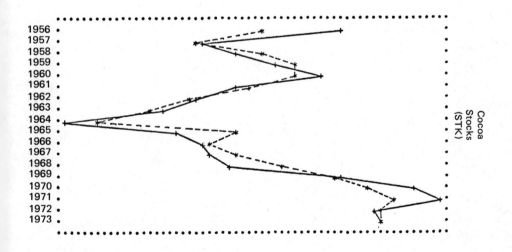

Cocoa
Stocks
(STK)

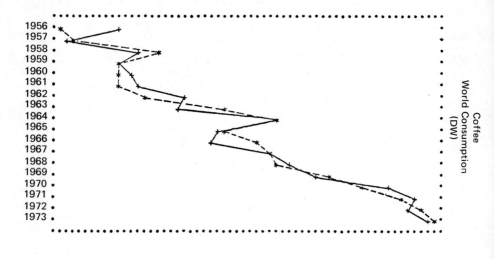

Coffee
World Consumption
(DW)

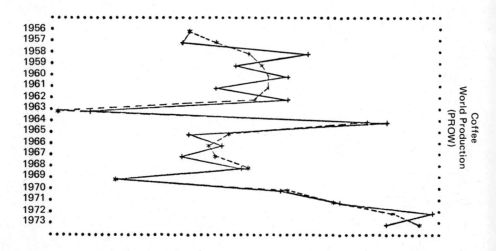

Coffee
World Production
(PROW)

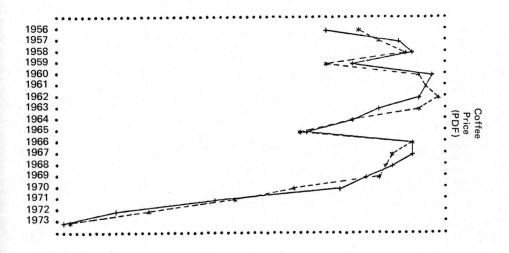

Coffee
Price
(PDF)

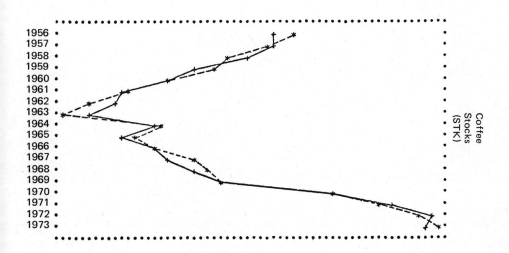

Coffee
Stocks
(STK)

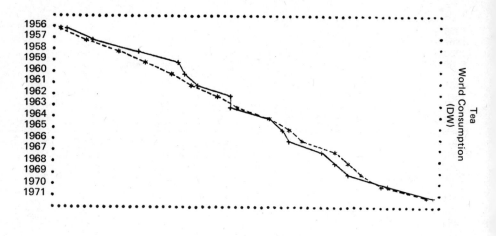

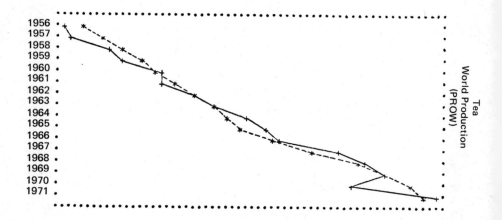

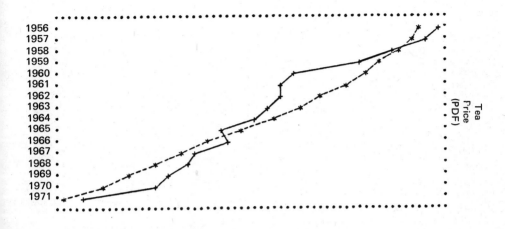

Tea
Price
(PDF)

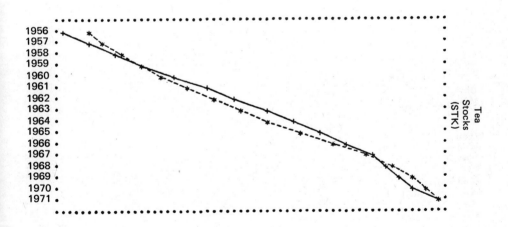

Tea
Stocks
(STK)

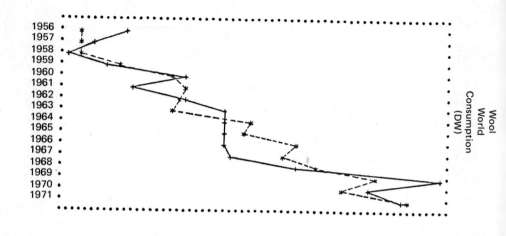

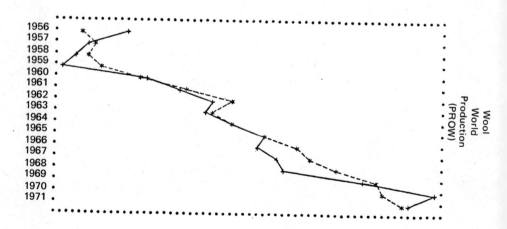

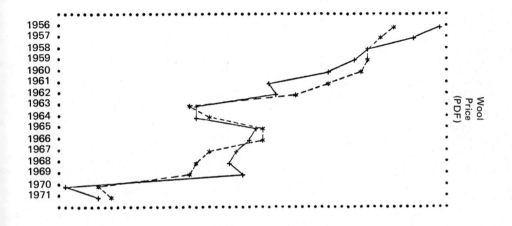

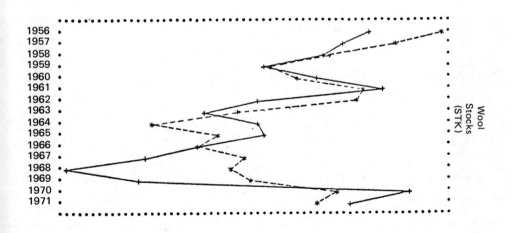

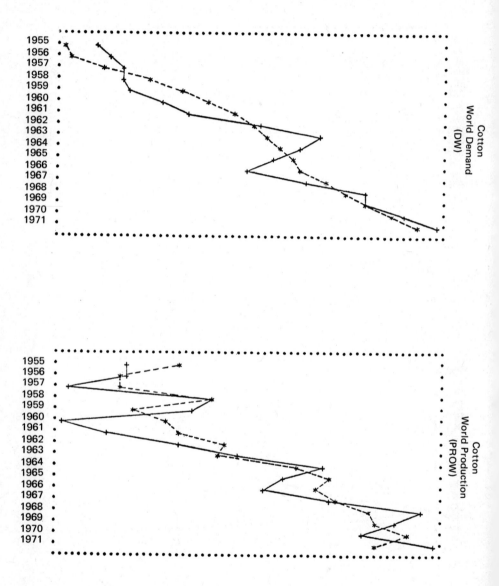

Cotton
World Demand
(DW)

Cotton
World Production
(PROW)

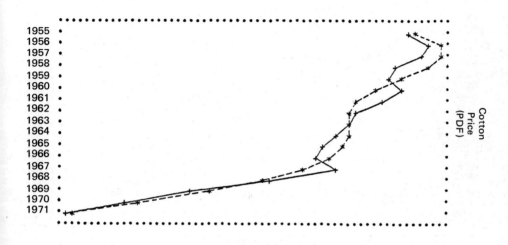

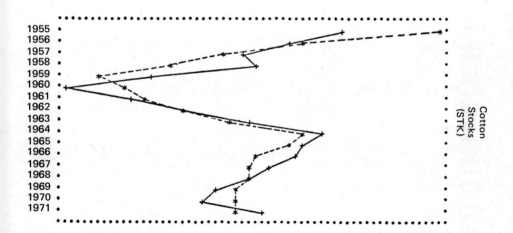

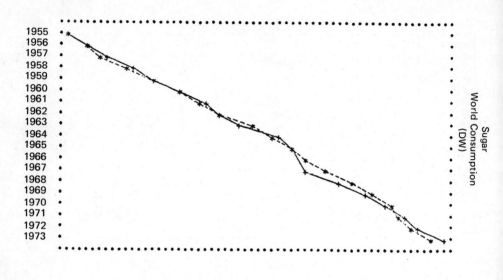

1955
1956
1957
1958
1959
1960
1961
1962
1963
1964
1965
1966
1967
1968
1969
1970
1971
1972
1973

Sugar
World Consumption
(DW)

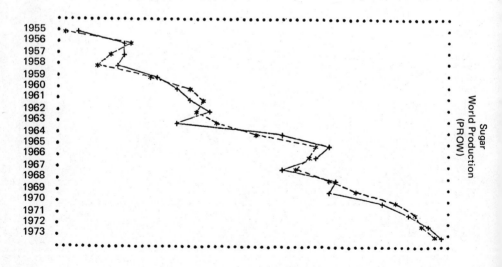

1955
1956
1957
1958
1959
1960
1961
1962
1963
1964
1965
1966
1967
1968
1969
1970
1971
1972
1973

Sugar
World Production
(PROW)

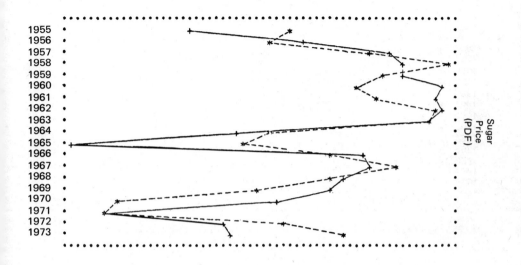

1955
1956
1957
1958
1959
1960
1961
1962
1963
1964
1965
1966
1967
1968
1969
1970
1971
1972
1973

Sugar
Price
(PDF)

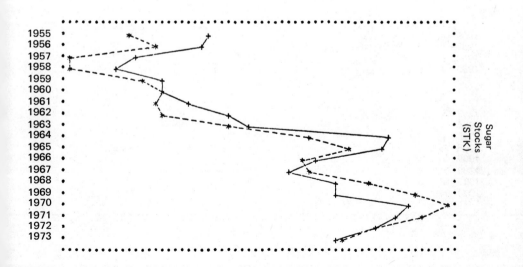

1955
1956
1957
1958
1959
1960
1961
1962
1963
1964
1965
1966
1967
1968
1969
1970
1971
1972
1973

Sugar
Stocks
(STK)

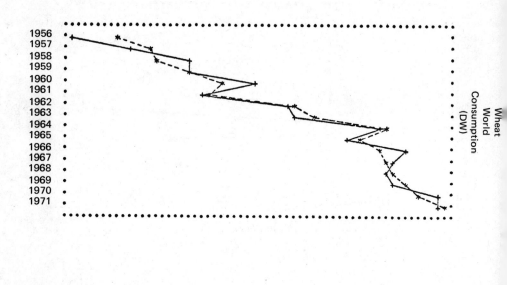

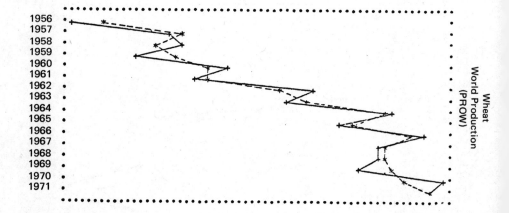

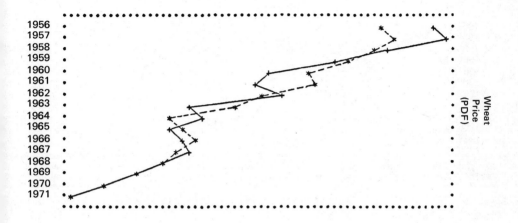

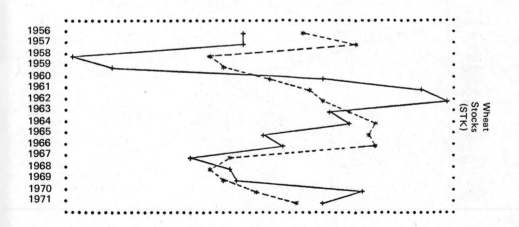

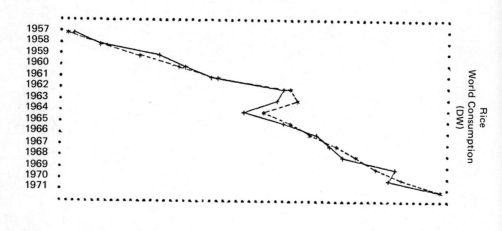

1957
1958
1959
1960
1961
1962
1963
1964
1965
1966
1967
1968
1969
1970
1971

Rice
World Consumption
(DW)

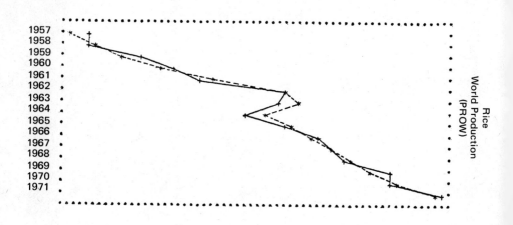

1957
1958
1959
1960
1961
1962
1963
1964
1965
1966
1967
1968
1969
1970
1971

Rice
World Production
(PROW)

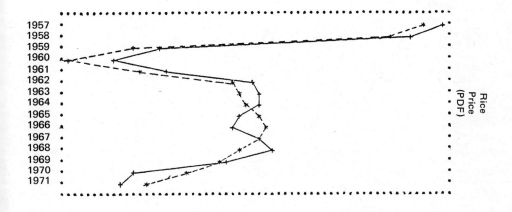

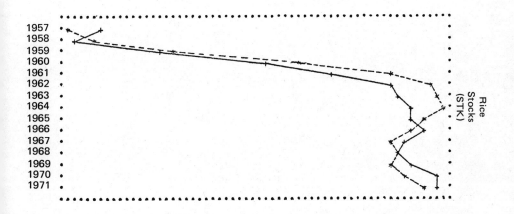

Bibliography

Acquah, Paul. 1972. "A Macroeconetric Analysis of Export Instability in Economic Growth: The Case of Ghana and the World Cocoa Market." Ph.D. dissertation, University of Pennsylvania, 1972.

Adams, F. Gerard. 1974. "The Impact of Nickel Production from the Ocean Floor: An Econometric Appraisal." Paper prepared for the United Nations Conference on Trade and Development, March 1974. Philadelphia: Economics Research Unit, 1974. Processed.

——. 1973a. "The Integration of World Primary Commodity Markets into Project LINK: The Example of Copper." Paper prepared for the annual meeting of Project LINK, Stockholm, September 1973. Philadelphia: Economics Research Unit, 1973. Processed.

——. 1973b. "The Impact of Copper Production from the Ocean Floor: Application of an Econometric Model." Paper prepared for the United Nations Conference on Trade and Development, December 1973. Philadelphia: Economics Research Unit, 1973. Processed.

——. 1972. "The Impact of Cobalt Production from the Ocean Floor: A Review of Present Empirical Knowledge and Preliminary Appraisal." Paper prepared for the United Nations Conference on Trade and Development, March 15, 1972. Philadelphia: Economics Research Unit, 1972. Processed.

Amoa, Rowland Ca-Kwami. 1965. "A Study in Demand: An Analysis of the Cocoa Bean and Cocoa Products Markets of the United States." Ph.D. thesis, Massachusetts Institute of Technology, 1965.

Arak, M. 1969. "Estimation of Assymetric Long-run Supply Functions: The Case of Coffee." *Canadian Journal of Agricultural Economics* (1969), pp. 15-21.

Arrow, Kenneth J. 1961. "Additive Logarithmic Demand Functions and the

Slutsky Relations." *Review of Economic Studies* 28 (June 1961): 166-181.

Bacha, E.L. 1968. "An Econometric Model for the World Coffee Market: The Impact of Brazilian Price Policy." Ph.D. dissertation, Yale University, 1968.

Ball, R.J., ed. 1973. *The International Linkage of National Econometric Models.* Amsterdam: North-Holland Publishing Co., 1973.

Bateman, Merrill J. 1965a. "Aggregate and Regional Supply Functions for Ghanaian Cocoa, 1946-1962." *J. Farm Econ.* 47 (May 1965): 384-401.

———. 1965b. "Cocoa in the Ghanaian Economy." Ph.D. thesis, Massachusetts Institute of Technology, 1965.

Behrman, Jere R. 1971. "Econometric Model Simulations of the World Rubber Market, 1950-1980." In Lawrence R. Klein, ed., *Essays in Industrial Econometrics,* vol. III. Philadelphia: University of Pennsylvania, Economics Research Unit, 1971.

———. 1968a. "Monopolistic Cocoa Pricing." *American Journal of Agricultural Economics* (1968), pp. 702-19.

———. 1968b. *Supply Response in Underdeveloped Agriculture: A Case Study of Four Major Annual Crops in Thailand,* 1937-1963. Amsterdam: North-Holland Publishers, 1968.

———. 1965. "Cocoa: A Study of Demand Elasticities in the Five Leading Consuming Countries, 1950-1961." *J. Farm Econ.* 47 (May 1965): 410-17.

Brown, C.P. 1975. *Primary Commodity Control.* Kuala Lumpur: Oxford University Press, 1975.

Duane, P. 1971. "Analysis of Wool Price Fluctuations: An Economic Analysis of Price Formation in a Raw Materials Market." Ph.D. dissertation, North Carolina State University, Raleigh, 1971.

Durbin, S.I. 1969. "A Sample Wool Marketing Model." MSc. thesis, Massey University, New Zealand, 1969.

Epps, Mary Lee S. 1970. "A Computer Simulation Model of the World Coffee Economy." Ph.D. dissertation, Duke University, 1970.

Goreux, Louis, M. 1972. "Price Stabilization Policies in World Markets for Primary Commodities: An Application to Cocoa." International Bank for Reconstruction and Development, January 1972. Mimeographed.

Gorman, William Moore. 1963. "Additive Logarithmic Preferences: A Further Note." *Review of Economic Studies* 30 (February 1963): 56-62.

Harberger, Arnold. 1963. "The Dynamics of Inflation in Chile." In Carl Christ, Ed., *Measurement in Economics: Studies in Mathematical Economics in Memory of Yehuda Grunfeld,* pp. 225-27. Stanford: Stanford University Press, 1963.

Holder, Shelby, Jr.; Shaw D.L.; and Snyder, J.C. 1970. *A Systems Model of the U.S. Rice Industry.* Technical Bulletin no. 1453. Washington, D.C.: U.S. Department of Agriculture, Economic Research Service, 1970.

Hoyt, R.C. 1955. "A Dynamic Econometric Model of the Milling and Baking Industries." Ph.D. dissertation, University of Minnesota, 1972. Meinken, K.W. *The Demand and Price Structure for Wheat.* Technical Bulletin No. 1136. United States Department of Agriculture, Washington, D.C., 1968.

Kofi, Tetteh A. 1972. "International Commodity Agreements and Export Earnings: Simulation of the 1968 Draft International Cocoa Agreement." *Food Research Institute Studies* 11 (1972).

Kolbe, H., and Timm, H. 1971. *Die Bestimmungsfaktoren der Preisentwicklung auf dem Weltmarkt fur Baumwolle: Eine Okonometrische Modellanalyse.* Hamburg: NR 4, HWWA-Instut fur Wirtschaftsforschung. July 1971.

Labys, Walter C. *Dynamic Commodity Models: Specification, Estimation and Simulation.* Lexington, Mass.: D.C. Heath, 1973.

——, ed. *Quantitative Models of Commodity Markets.* Cambridge, Mass.: Ballinger, 1975.

Law, Alton D. 1975. *International Commodity Agreements.* Lexington, Mass.: D.C. Heath, 1975.

Mathis, Kary. 1969. *An Economic Situation Model of the Cacao Industry of the Dominican Republic.* International Programs Info. Rep. No. 69-2. Texas A & M University, Department of Agricultural Economics and Sociology, 1969.

McKenzie, C.J. 1966. "Quarterly Models of Price Formation in the Raw Wool Market." M.Sc. thesis, Lincoln College, Canterbury, New Zealand, 1966.

Mo, Y. 1968. *An Economic Analysis of the Dynamics of the United States Wheat Sector.* Technical Bulletin No. 1395. Washington, D.C.: U.S. Department of Agriculture, 1968.

Nasol, R.L. 1971. "Demand Analysis for Rice in the Philippines." *Journal of Agricultural Economics and Development* (1971), pp. 1–13.

Nerlove, Marc. 1958. *The Dynamics of Supply: Estimations of Farmers' Response to Price.* Baltimore: The Johns Hopkins University Press, 1958.

Oury, Bernard. 1966. *A Production Model For Wheat and Feedgrains in France, (1946-1961).* Amsterdam: North-Holland Publishing Co., 1966.

Shayal, S.E.M. 1960. "An Econometric Model of the Egyptian Cotton Market." Dissertation, Oxford University, 1961.

Tewes, T. 1972. "Sugar, A Short Term Forecasting Model for the World Market." *The Business Economist,* (Summer 1972), pp. 89–97.

Tsujii, Hiroshi. 1973. "An Econometric Study of Effects of National Rice Policies and the Grain Revolution on National Rice Economies. . . ." Ph.D. dissertation, University of Illinois, Urbana, 1973.

United Nations. 1970. *Methods Used in Compiling the United Nations Price Indexes for Basic Commodities in International Trade.* Statistical Papers, series M, no. 29, rev. 7. New York, 1970.

United Nations, FAO. 1965. "Cocoa: The Outlook for Production and Consumption, 1970-1972." Working Party-Prices and Quotas, for UN Cocoa Conference, New York, 1965. Mimeographed.

Vannerson, F.L. 1969. "An Econometric Analysis of the Postwar U.S. Wheat Market." Ph.D. dissertation, Princeton University, 1969.

Weymar, F. Helmut. 1968. *The Dynamics of the World Cocoa Market.* Cambridge, Mass.: The MIT Press, 1968.

Wickens, M.; Greenfield, J.; and Marshall, G. 1971. *A World Coffee Model.*

CCP.71/W.P.4.. Rome: Food and Agricultural Organization of the United Nations, 1971.

Wickens, M.R. and Greenfield, J.N. "The Econometrics of Agricultural Supply: An Application to the World Coffee Market." *Review of Economics and Statistics* 55(4) (November 1973): 433–40.

Wilson, John. 1973. "An Econometric Analysis of Brazilian Coffee Supply." University of Pennsylvania, Philadelphia, 1973. Mimeographed.

Witherell, William H. 1967. *Dynamics of the International Wool Market: An Econometric Analysis.* Research Memorandum no. 91. Princeton, N.J.: Princeton University, Econometric Research Program, 1967.

Wymer, C.R. 1975. "Estimation of Continuous Time Models with an Application to the World Sugar Market." In Walter C. Labys, ed., *Quantitative Models of Commodity Markets,* pp. 173-91. Cambridge, Mass.: Ballinger, 1975.

About the Authors

F. Gerard Adams is Professor of Economics and Finance at the University of Pennsylvania and heads its Economics Research Unit. He is also Treasurer and Senior Consultant to Wharton EFA, Inc. He has combined an academic career with extensive consultation to business, the U.S. government, and international organizations. His work in applied econometrics has ranged widely from model building and forecasting for the nation to studies of industries, regions, and commodity markets. These studies have included an econometric model of international trade, an econometric-linear programming model of petroleum refining, empirical studies of income distribution and consumer and investor anticipations, macroeconomic model building for the U.S., Belgium and Brazil, and numerous commodity market models. He has published widely.

Jere R. Behrman is Chairperson of the Economics Department and Professor of Economics at the University of Pennsylvania. He is the author of several books on agricultural supply response, economic development, international trade, commodity markets, and intergenerational transfer of income and status. He has also contributed articles to such journals as the *American Economic Review, Econometrica, Journal of Political Economy,* and the *International Economic Review.* He has served as a consultant for Wharton EFA, the Brookings Institution, the World Bank, the Harvard Institute for International Development, UNCTAD and the UNDP.